T0297587

SEDIMENT DYNAMICS OF CHINESE MUDDY COASTS AND ESTUARIES

SEDIMENT DYNAMICS OF CHINESE MUDDY COASTS AND ESTUARIES

Physics, Biology, and Their Interactions

Edited by

XIAO HUA WANG

Academic Press is an imprint of Elsevier
125 London Wall, London EC2Y 5AS, United Kingdom
525 B Street, Suite 1650, San Diego, CA 92101, United States
50 Hampshire Street, 5th Floor, Cambridge, MA 02139, United States
The Boulevard, Langford Lane, Kidlington, Oxford OX5 1GB, United Kingdom

Notices
Knowledge and best practice in this field are constantly changing. As new research and experience broaden our
understanding, changes in research methods, professional practices, or medical treatment may become necessary.

Practitioners and researchers must always rely on their own experience and knowledge in evaluating and using
any information, methods, compounds, or experiments described herein. In using such information or methods
they should be mindful of their own safety and the safety of others, including parties for whom they have a
professional responsibility.

To the fullest extent of the law, neither the Publisher nor the authors, contributors, or editors, assume any liability
for any injury and/or damage to persons or property as a matter of products liability, negligence or otherwise, or
from any use or operation of any methods, products, instructions, or ideas contained in the material herein.

Library of Congress Cataloging-in-Publication Data
A catalog record for this book is available from the Library of Congress

British Library Cataloguing-in-Publication Data
A catalogue record for this book is available from the British Library

ISBN: 978-0-12-811977-8

For information on all Academic Press publications visit our
website at https://www.elsevier.com/books-and-journals

Working together
to grow libraries in
developing countries

www.elsevier.com • www.bookaid.org

Publisher: Candice Janco
Acquisition Editor: Louisa Hutchins
Editorial Project Manager: Hilary Carr
Production Project Manager: Maria Bernard
Cover Designer: Matthew Limbert

Typeset by SPi Global, India

CONTENTS

Contributors ix

Chapter 1 Introduction **1**
Isabel Jalón-Rojas, Xiao Hua Wang

References 4

Chapter 2 Jiaozhou Bay **5**
Lulu Qiao, Shengkang Liang, Dehai Song, Wen Wu, Xiao Hua Wang

1 Background 6
2 Environmental Responses to Human Interventions 12
3 Environmental Management 17
4 Conclusions 19
References 20

**Chapter 3 Muddy Coast Off Jiangsu, China: Physical, Ecological,
and Anthropogenic Processes** **25**
Jiabi Du, Benwei Shi, Jiasheng Li, Ya Ping Wang

1 Introduction 26
2 Physical Control 27
3 Ecological Control 34
4 Human Intervention 37
5 Weakness of Current Studies on Jiangsu Mudflat 41
6 Summary 44
Acknowledgments 45
References 45
Further Reading 49

Chapter 4   Changjiang Estuary 51
Jianrong Zhu, Hui Wu, Lu Li, Cheng Qiu

1 Introduction 51
2 Numerical Model 53
3 Dynamic Factors Controlling the Water Movement in the
  Changjiang Estuary 55
4 Saltwater Intrusion 62
5 Hypoxia off the River Mouth 67
6 Summary 71
References 73
Further Reading 75

Chapter 5   Changes in the Hydrodynamics of Hangzhou Bay Due
  to Land Reclamation in the Past 60 Years 77
Li Li, Taoyan Ye, Xiao Hua Wang, Zhiguo He, Ming Shao

1 Introduction 78
2 Urbanization 78
3 Changes of Physical Environment 82
4 Discussion 90
5 Conclusions 91
Acknowledgments 91
References 91

Chapter 6   Marine Environmental Status and Blue Bay
  Remediation in Xiamen 95
Keliang Chen, Senyang Xie, Hongzhe Chen

1 Climatology, Hydrology, and Geology .95
2 Biodiversity and Ecological Disasters .100
3 Human Interventions: Reclamation and Its Environmental
  Impacts 104

4 Modeling Approach 108

5 Recommendations and Future Steps 117

References 120

Further Reading 122

Chapter 7 Coastal Dynamics and Sediment Resuspension in Laizhou Bay **123**

Zai-Jin You, Chao Chen

1 Laizhou Bay 123

2 Meteorological Climate 125

3 Coastal Dynamics 127

4 Sediment Transport 132

5 Human Impact 137

References 140

Further Reading 141

Chapter 8 Remarks **143**

Xiao Hua Wang, Isabel Jalón-Rojas

Index 147

CONTRIBUTORS

Chao Chen School of Engineering and Technology, Jimei University, Xiamen, China

Hongzhe Chen Third Institute of Oceanography, State Oceanic Administration, Xiamen, China

Keliang Chen Third Institute of Oceanography, State Oceanic Administration, Xiamen, China

Jiabi Du Virginia Institute of Marine Science, College of William and Mary, Gloucester Point, VA; Department of Marine Sciences, Texas A&M University, Galveston, TX, USA

Zhiguo He Ocean College, Zhejiang University, Zhoushan; State Key Laboratory of Satellite Ocean Environment Dynamics, Second Institute of Oceanography, Hangzhou, China

Isabel Jalón-Rojas The Sino-Australian Research Centre for Coastal Management, The University of New South Wales, Canberra, ACT, Australia

Jiasheng Li Ministry of Education Key Laboratory for Coast and Island Development, Nanjing University, Nanjing, China

Lu Li State Key Laboratory of Estuarine and Coastal Research, East China Normal University, Shanghai, China

Li Li Ocean College, Zhejiang University, Zhoushan; State Key Laboratory of Satellite Ocean Environment Dynamics, Second Institute of Oceanography, Hangzhou, China

Shengkang Liang Key Laboratory of Marine Chemistry Theory and Technology, Ministry of Education, Qingdao, China

Lulu Qiao Key Lab of Sea Floor Resource and Exploration Technique, Ministry of Education; College of Marine Geosciences, Ocean University of China, Qingdao, China

Cheng Qiu State Key Laboratory of Estuarine and Coastal Research, East China Normal University, Shanghai, China

Ming Shao Ocean College, Zhejiang University, Zhoushan, China

Benwei Shi Ministry of Education Key Laboratory for Coast and Island Development, Nanjing University, Nanjing, China

Dehai Song   Key Laboratory of Physical Oceanography, Ministry of Education, Ocean University of China, Qingdao, China

Xiao Hua Wang   State Key Laboratory of Satellite Ocean Environment Dynamics, Second Institute of Oceanography, Hangzhou, China; The Sino-Australian Research Centre for Coastal Management, The University of New South Wales, Canberra, ACT, Australia

Ya Ping Wang   Ministry of Education Key Laboratory for Coast and Island Development, Nanjing University, Nanjing; State Key Laboratory of Estuarine and Coastal Research, East China Normal University, Shanghai, China

Wen Wu   College of Oceanic and Atmospheric Sciences, Ocean University of China, Qingdao, China

Hui Wu   State Key Laboratory of Estuarine and Coastal Research, East China Normal University, Shanghai, China

Senyang Xie   School of Physical, Environmental, and Mathematical Sciences; The Sino-Australian Research Centre for Coastal Management, University of New South Wales, Canberra, ACT, Australia

Taoyan Ye   Ocean College, Zhejiang University, Zhoushan, China

Zai-Jin You   Ports and Coastal Research Centre, Ludong University, Yantai, China

Jianrong Zhu   State Key Laboratory of Estuarine and Coastal Research, East China Normal University, Shanghai, China

INTRODUCTION

Isabel Jalón-Rojas, Xiao Hua Wang

The Sino-Australian Research Centre for Coastal Management, The University of New South Wales, Canberra, ACT, Australia

China has experienced an unprecedented economic growth since the beginning of market-oriented reforms and opening-up in 1978. China's rapidly growing industry and trade were particularly concentrated in coastal regions, which support almost half of China's population (Kanbur and Zhang, 2005; Holz, 2008). Proof of this is the coastal gross domestic product increase in more than two orders of magnitude between 1978 and 2010 (He et al., 2014).

This rapid development has implied significant pressures on China's coastal ecosystems. Human pressures include: (1) land reclamation for agriculture and urban expansion (Wang et al., 2014; Gao et al. 2014); (2) construction of infrastructure projects such as harbors and bridges (Wang and Slack, 2004; Li et al., 2014); (3) increased runoff of nutrients and contaminants (Pan and Wang, 2012); (4) introduction of invasive species (An et al., 2007); (5) industrialized overfishing (Coulter, 1996); and (6) change of the hydrological regime by freshwater abstractions and dam construction (Yang et al., 2008). These pressures can lead to dramatic changes of the physical environment of coastal systems and consequently to the degradation and loss of ecosystem services and habitats (Wang et al., 2017).

Disturbance and change in estuarine and coastal systems also occur in response to global climate change (Day et al., 2012; Robins et al., 2016). Coastal zones are prone to patterns changes of climate processes such as sea-level rise, wind, precipitation, surface heat budget (i.e., temperature, solar radiation, evaporation) and ocean acidification (IPCC, 2014). Anthropic pressures interact with each other and with climate stresses, leading to cumulative effects and complex feedbacks on coastal bio-physical processes, which are not yet fully understood.

Sediment Dynamics of Chinese Muddy Coasts and Estuaries. https://doi.org/10.1016/B978-0-12-811977-8.00001-7

This context of coastal degradation has globally given rise to social, economic, and political concerns (Gössling and Hall, 2005). Understanding the impact of human activities on coastal physical environments is thus critical to establish a path for economic development with a healthy and sustainable environment. This will allow evaluation of what we have learned from past experiences, rethinking of traditional coastal engineering approaches, such as constructing breakwaters and other rigid structures that have environmental impacts, and the proposition of future sustainable management strategies. But we need to address these issues, not just locally, but at the largest regional scales, and also to consider the decadal (and longer) timescales over which the environment responds to human alterations and global climate change (Little et al., 2017).

The multiple human interventions along the China's 30,017 km of coastline offer an ideal context to respond to these challenging problems. With this in mind, this book aims to examine and document the physical and ecological impacts on Chinese estuarine and coastal environment due to intensive human activities and their effects on the coastal ecosystems. The book reviews and synthesizes the latest research on physical, sedimentary, biological, chemical, and ecological processes associated with rapidly changing environments, as well as its interactions (Fig. 1). In brief, human-induced morphological and hydrological changes lead to the alteration of hydrodynamics, and therefore, of sediment transport, nutrient and pollutant transport, and mixing processes. Biological and chemical processes determine the response of aquatic communities to these perturbations. Ecosystem communities can also be directly perturbed by morphological changes (e.g., losses of salt marshes and mangroves) and by the introduction of new species which can in turn modify sediment dynamics, such as deposition rates, and therefore, morphology.

By compiling the latest results of leading and interdisciplinary scientists concerned with Chinese coastal processes, this book deepens in these complex biophysicochemical interactions using different and complementary observational and modeling approaches and six study cases: Jiaozshu Bay, Jiangshu Coast, Hangzhou Bay, Changjiang Estuary, Xiamen Bay, and Laizhou Bay (Fig. 2). The final goal is to provide a forum to build possible solutions to coastal degradation and recommendations for an ecologically sustainable development that can be transferable on the world stage.

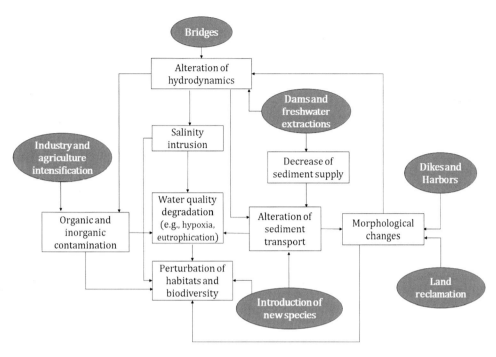

Fig. 1 Scheme of the interactions between biophysiochemical processes *(white rectangles)* associated to human interventions *(shadow ovals)*.

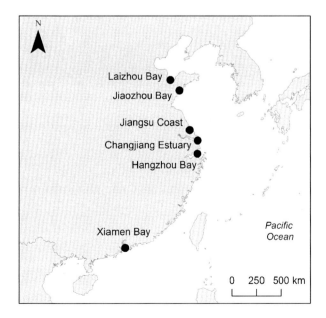

Fig. 2 Location map of study sites.

References

An, S.Q., Gu, B.H., Zhou, C.F., Wang, Z.S., Deng, Z.F., Zhi, Y.B., Liu, Y.H., 2007. Spartina invasion in China: implications for invasive species management and future research. Weed Res. 47 (3), 183–191. https://doi.org/10.1111/j.13653180.2007.00559.x.

Coulter, D.Y., 1996. South China Sea fisheries: countdown to calamity. Contemp. Southeast Asia 17 (4), 371–388.

Day, J.W., Yáñez-Arancibia, A., Rybczyk, J.M., 2012. Climate change: effects, causes, consequences: physical, hydromorphological, ecophysiological, and biogeographical changes. In: Treatise on Estuarine and Coastal Science. vol. 8, pp. 303–315. https://doi.org/10.1016/B978-0-12-374711-2.00815-9.

Gao, G.D., Wang, X.H., Bao, X.W., 2014. Land reclamation and its impact on tidal dynamics in Jiaozhou Bay, Qingdao, China. Estuar. Coast. Shelf Sci. 151, 285–294. https://doi.org/10.1016/j.ecss.2014.07.017.

Gössling, S., Hall, C.M. (Eds.), 2005. Tourism and Global Environmental Change: Ecological, Social, Economic and Political Interrelationships. Routledge, London, New York. https://doi.org/10.4324/9780203011911.

He, Q., et al., 2014. Economic development and coastal ecosystem change in China. Scientific Reports 4, 5995. https://doi.org/10.1038/srep 05995.

Holz, C.A., 2008. China's economic growth 1978-2025: what we know today about China's economic growth tomorrow. World Dev. 36 (10), 1665–1691. https://doi.org/10.1016/j.worlddev.2007.09.013.

IPCC, 2014. In: Climate change 2014: impacts, adaptation, and vulnerability. Part B: Regional aspects. Contribution of Working Group II to the Fifth Assessment Report of the Intergovernmental Panel on Climate Change. Cambridge University Press, p. 688. https://doi.org/10.1017/CBO9781107415324.004.

Kanbur, R., Zhang, X., 2005. Fifty years of regional inequality in China: a journey through central planning, reform, and openness. Rev. Dev. Econ. 9 (1), 87–106. https://doi.org/10.1111/j.1467-9361.2005.00265.x.

Li, P., Li, G., Qiao, L., Chen, X., Shi, J., Gao, F., Yue, S., 2014. Modeling the tidal dynamic changes induced by the bridge in Jiaozhou Bay, Qingdao, China. Cont. Shelf Res. 84, 43–53. https://doi.org/10.1016/j.csr.2014.05.006.

Little, S., Spencer, K.L., Schuttelaars, H.M., Millward, G.E., Elliott, M., 2017. Unbounded boundaries and shifting baselines: estuaries and coastal seas in a rapidly changing world. Estuar. Coast. Shelf Sci. 198, 311–319. https://doi.org/10.1016/j.ecss.2017.10.010.

Pan, K., Wang, W.-X., 2012. Trace metal contamination in estuarine and coastal environments in China. Sci. Total Environ. 421–422, 3–16. https://doi.org/10.1016/j.scitotenv.2011.03.013.

Robins, P.E., Skov, M.W., Lewis, M.J., Giménez, L., Davies, A.G., Malham, S.K., Jago, C.F., 2016. Impact of climate change on UK estuaries: a review of past trends and potential projections. Estuar. Coast. Shelf Sci. https://doi.org/10.1016/j.ecss.2015.12.016.

Wang, J.J., Slack, B., 2004. Regional governance of port development in China: a case study of Shanghai International Shipping Center. Marit. Policy Manag. 31 (4), 357–373. https://doi.org/10.1080/0308883042000304467.

Wang, W., Liu, H., Li, Y., Su, J., 2014. Development and management of land reclamation in China. Ocean Coast. Manag. 102, 415–425. https://doi.org/10.1016/j.ocecoaman.2014.03.009.

Wang, X.H., Gan, J., Lowe, R., 2017. Sediment dynamics of muddy coasts and estuaries in China: an introduction. Estuar. Coast. Shelf Sci.. https://doi.org/10.1016/J.ECSS.2017.11.036.

Yang, T., Zhang, Q., Chen, Y.D., Tao, X., Xu, C.y., Chen, X., 2008. A spatial assessment of hydrologic alteration caused by dam construction in the middle and lower Yellow River, China. Hydrol. Process. 22 (18), 3829–3843. https://doi.org/10.1002/hyp.6993.

2

JIAOZHOU BAY

Lulu Qiao*,†, Shengkang Liang‡, Dehai Song§, Wen Wu¶, Xiao Hua Wang‖

Key Lab of Sea Floor Resource and Exploration Technique, Ministry of Education, Qingdao, China †College of Marine Geosciences, Ocean University of China, Qingdao, China ‡Key Laboratory of Marine Chemistry Theory and Technology, Ministry of Education, Qingdao, China §Key Laboratory of Physical Oceanography, Ministry of Education, Ocean University of China, Qingdao, China ¶College of Oceanic and Atmospheric Sciences, Ocean University of China, Qingdao, China ‖The Sino-Australian Research Centre for Coastal Management, The University of New South Wales, Canberra, ACT, Australia

CHAPTER OUTLINE
1 **Background   6**
    1.1 Coastline and Morphology   6
    1.2 Hydrodynamics   6
    1.3 Sediment Transport   8
    1.4 Water Quality   9
    1.5 Biology   11
2 **Environmental Responses to Human Interventions   12**
    2.1 Coastline Change   12
    2.2 Hydrodynamics   13
    2.3 Sediment Transport   13
    2.4 Water Quality   14
    2.5 Biology   15
3 **Environmental Management   17**
    3.1 Case Study: Cross-Sea Bridge   17
    3.2 Case Study: Hongdao Channel   18
    3.3. Politics and Laws   18
4 **Conclusions   19**
References   20

Sediment Dynamics of Chinese Muddy Coasts and Estuaries. https://doi.org/10.1016/B978-0-12-811977-8.00002-9

Jiaozhou Bay (JZB) is a semi-enclosed shallow water situated off the southern coast of the Shandong Peninsula, China (Fig. 1A). The bay was transformed from a continental basin to a coastal bay by the rising sea level since the end of the Last Glacial period (Yan et al., 2000). Thus, the JZB is about ten thousand years old (Li et al., 2014a,b). Now, the stable crust in this area indicates that the current environmental evolution of JZB is mostly controlled by modern hydrodynamics, human activities, and even climate change.

1 Background

1.1 Coastline and Morphology

The total length of the JZB coastline in 2009 was 178.06 km, of which 158.81 km was artificial, and 19.25 km was natural, accounting for about 89.19% and 10.81% of the total length, respectively (Zhang, 2009). Large area of natural tidal flat was replaced by salt pond, aquaculture area, and reclamation for ports (Zhou et al., 2010), so the coastline tends to be straight, especially along the bay entrance and its eastern coast. Furthermore, seaward reclamation around the Yanghe River mouth extended the coastline, the total length of which increased to 187.3 km in 2013.

The surface area of JZB was 339.3 km^2 in 2013, with an average water depth of 7 m (Gao and Wang, 2002). Large areas of tidal flat and shallow water are located in northwestern JZB (Fig. 1A), which is significant to water exchange and environmental capacity. A deep trough over 64 m can be found in the southeast (Bai, 2005).

The seabed morphology of JZB can be divided into two types: accumulation and abrasion geomorphy. Accumulation geomorphy, exhibited as tidal sand ridges, intertidal shoals, accumulation plains, and subaqueous deltas, can be found mostly in northern JZB (Li et al., 1986). Abrasion geomorphy, characterized by four finger-like scour troughs, are mainly in central JZB, which are the path of tidal currents (Li et al., 2014a,b). One of the deep troughs, the Cangkou Channel, connects to the deep water at the bay mouth and is the most important ship route in Qingdao.

1.2 Hydrodynamics

JZB has a semi-diurnal tidal regime, that is, it is dominated by flood tidal currents. The mean tidal range is 2.78 m with a maximum of 4.75 m measured at Dagang tide-gauge station inside the Qingdao Port (Fig. 1A). Due to the complex coastline and bathymetry, including headlands, islands, reefs, and deep

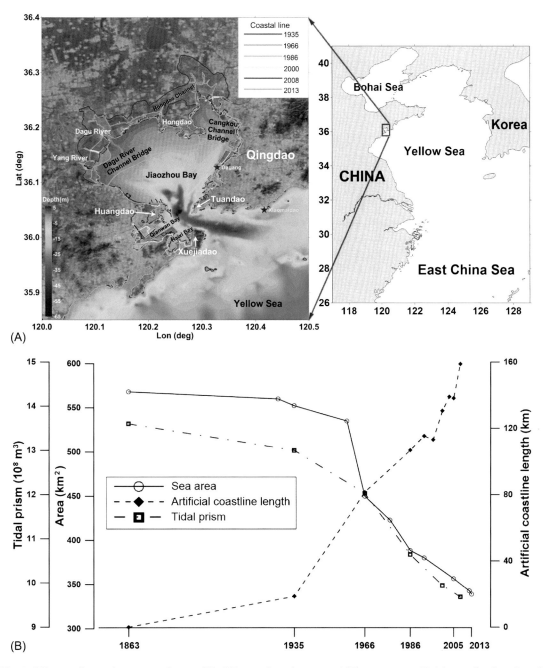

Fig. 1 JZB coastline and sea area change. (A) JZB coastline change; and (B) sea area, artificial coastline length and tidal prism change over the period of 1966 to 2013. Data from Zhou, C.Y., Li, G.X., Shi, J.H., 2010. Coastline change of Jiaozhou Bay over the last 150 years. Period. Ocean Univ. China 40, 99–106, (in Chinese with English abstract); Shi, J., Li, G., Wang, P., 2011. Anthropogenic influences on the tidal prism and water exchanges in Jiaozhou Bay, Qingdao, China. J. Coast. Res. 27, 57–72; Chen, J., Chen, X., 2012. Numerical simulation of the hydrodynamic evolution of the Jiaozhou Bay in the last 70 years. Acta Oceanol. Sin. 34, 30–41 (in Chinese with English abstract); Ma, L.J., Yang, X.G., Qi, Y.L., Liu, Y.X., Zhang, J.Z., 2014. Oceanicarea change and contributing factor of Jiaozhou Bay. Sci. Geogr. Sin. 3, 365–369 (in Chinese with English abstract).

channels, strong currents over $300\,\mathrm{cm\,s^{-1}}$ can be found in this region. The strong, turbulent mixing caused by tides generates the vertically homogenized profiles of temperature, salinity, and biochemical variables throughout the year (Weng et al., 1992).

JZB has the earliest wave observation station (Xiaomaidao Station, Fig. 1A) in China. The waves in JZB are mainly controlled by monsoon, which generates northwestward-propagating waves in spring and summer, and southeastward-propagating waves in autumn and winter. As nearly 90% of the waves have a height less than 0.5 m, the waves in JZB are rather weak. However, during the storm that occurred during 16–20 August 1985, the south coastline of Qingdao suffered from a highest one-tenth wave height of $H_{1/10}=9\,\mathrm{m}$; meanwhile, only $H_{1/10}=2.8\,\mathrm{m}$ was measured near Huangdao Island (China's Harbors and Embayments, 1993), which indicates JZB is a well-sheltered harbor.

There are more than 13 rivers that discharge into JZB with freshwater and sediments, most of which are small seasonal rivers, and some have been turned into canals for industrial and domestic waste discharge. The Dagu River, which has a length of 179.9 km and a basin area of $6131.3\,\mathrm{km^2}$, is the largest river terminating in JZB (Sheng et al., 2014). The Dagu River has an annual water discharge of $7.235\times10^8\,\mathrm{m^3}$ (Jin et al., 2010), and an annual sediment discharge of $3.659\times10^5\,\mathrm{t}$ between 1960 and 2008 (Sheng et al., 2014), which has been reduced sharply in recent years (Zhao, 2016). It is also a seasonal river, with 89.8% of the total freshwater discharged into the bay during summer (Jiang and Wang, 2013).

1.3 Sediment Transport

JZB is recognized as a large tidal inlet with lower suspended sediment concentration (SSC), less sediment supply, and lower deposition rate (Gao and Wang, 2002). The SSC of JZB is lower than that in other coastal areas of the east Chinese seas (Wang et al., 2014), decreasing from the northwest with a value of $10–50\,\mathrm{mg\,L^{-1}}$ to the southeast with a value of less than $10\,\mathrm{mg\,L^{-1}}$ (Wang et al., 2014). The SSC reached its maximum value in spring, while its minimum value was in summer (Zhang, 2000). Because the sediment amount discharged by modern rivers is relatively small, the major source of the suspended sediment in JZB is resuspension, which is controlled by tidal currents and wind-driven waves. According to field observations, the net suspended sediments were transported within the bay entrance channel directed toward the open sea, with the flux of $10^6\,\mathrm{t\cdot a^{-1}}$ (Gao and Wang, 2002).

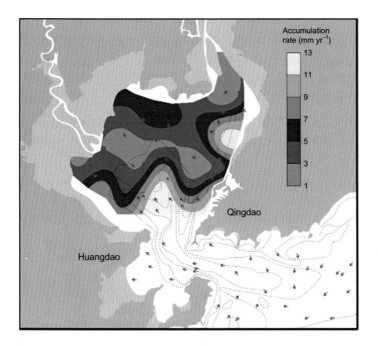

Fig. 2 Suspended sediment accumulation rates (colored in mm/a), net sediment transport pathway *(red arrows)* and bathymetry *(dashed lines)* in JZB. From Wang, Y.P., Gao, S., Jia, J.J., Liu, Y.L., Gao, J.H., 2014. Remarked morphological change in a large tidal inlet with low sediment-supply. Cont. Shelf Res. 90, 79–95.

JZB is a slowly silted bay with a low accumulation rate of 10^0 to 10^1 mm·a^{-1} in the centennial time scale (Fig. 2; Gao and Wang, 2002; Dai et al., 2007; Wang et al., 2014). A depocenter was identified at the flood delta (Bai, 2005; Wang et al., 2014). The seabed sediment in JZB becomes coarser from the northwest to southeast, with clay silt distributed widest (Yan et al., 2000; Wang et al., 2000).

1.4 Water Quality

Water quality is a fundamental indicator of aquatic ecological health and sustainability. In recent years, although the seawater quality of JZB was good as a whole, the deterioration of its seawater quality was still serious (Ocean & Fishery Administration of Qingdao, 2002–2015). The ratio of JZB water area with water quality exceeding the second level of the Chinese National Seawater Quality Standard (CNSQS) was maintained at about 40% (Ocean & Fishery Administration of Qingdao, 2002–2015). Major polluted areas were distributed in northern and eastern parts of JZB and in Haixi-Qianwan Bay (Ocean & Fishery Administration of Qingdao, 2002–2015).

The major environmental issue in JZB is eutrophication, due to excessive input of land-based wastewater containing high concentrations of inorganic dissolved nitrogen (DIN) and active

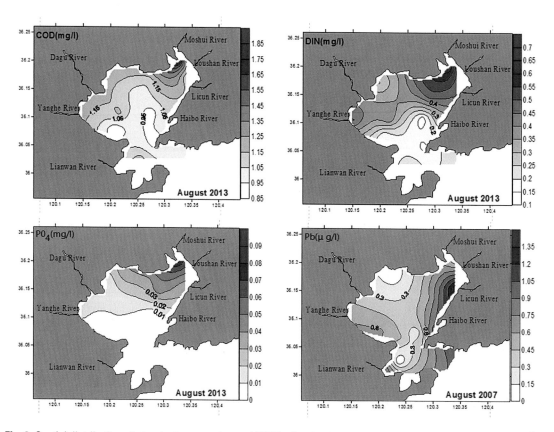

Fig. 3 Spatial distribution of chemical oxygen demand (COD), dissolved inorganic nitrogen, active phosphorus (PO_4^{3-}), and heavy metal (Pb) in the surface water of JZB in August of 2007 and 2013, respectively.

phosphate (PO_4^{3-}). The concentrations of DIN and PO_4^{3-} in JZB gradually decreased from the north to the central and bay mouth (Fig. 3), because several rivers with vast wastewater from agricultural, industrial, and domestic waste flow into the bay from the north (Ocean & Fishery Administration of Qingdao, 2002–2015). The fluxes of DIN and $PO4^-$ discharged into JZB are over the threshold of its environmental capacity (Han et al., 2011), and the demand on its ability to regulate pollutants is excessive.

Heavy metal contamination in the coastal water has become a major concern in terms of environmental safety. In JZB, Cd, Cu, Hg, and Pb are the primary heavy metal elements in water, among which Hg and Pb were often reported exceeding the first levels of the CNSQS (0.05 and 1.0 μg/L, respectively) (Ocean & Fishery Administration of Qingdao, 2002–2015). The heavy metals in JZB are mainly derived from atmospheric deposition, runoff, and nearshore pollution. Furthermore, the heavy metal input from rivers has great impact on its distribution in JZB (Wang et al., 2015), meaning that high concentrations of heavy metals usually presented near sewage outlets and river estuaries (Fig. 3).

Petroleum is another anthropogenic main pollutant in JZB. On average, more than one oil-spill event happened in Qingdao coastal area each year between 2009 and 2014 (Ocean & Fishery Administration of Qingdao, 2002–2015). An oil pipeline exploded in Huangdao on November 22, 2013, resulting in a mass of crude oil entering JZB and polluting the seawater in the bay. In fact, Qingdao Port is China's largest import point for crude oil, and Huangdao Port seeks to be an oil products and chemical logistics hub; thus, oil spills have been recognized as posing a great threat to JZB's environment and ecology.

The contamination level of persistent organic pollutants (POPs) in the seawater of JZB is relatively low at the present time in comparison with other industrial-affected areas worldwide (Fu et al., 2007; Xu et al., 2007; Zhang et al., 2016). Though the concentration of organochlorine (OCPs) is within the regulatory limits, for example, maximum residue limits for seawater as recommended by the United States Environmental Protection Agency (US EPA) and State Environmental Protection Administration of China (CSEPA) (Xu et al., 2007), high concentrations of nonylphenol (NP), an endocrine disruptor, was observed in the northeast and northwest parts of JZB, which poses a significant risk for aquatic ecosystems (Fu et al., 2007; Zhang et al., 2016).

1.5 Biology

JZB is regarded as a eutrophic bay with a high primary productivity (Sun et al., 2011a). In 2013, the average density of phytoplankton was 1.31×10^7 cell·m^{-3}, with an average biodiversity index of 2.29. A total of 95 species of phytoplankton was obtained, subjected to 35 genera of 23 families of 15 orders of five classes of three phyla, in which 72 species belonged to diatom, 21 species belonged to dinoflagellates, and two species belonged to chrysophyceae (Ocean & Fishery Administration of Qingdao, 2002–2015). A total of 48 species of zooplankton was obtained (fish eggs and larvae excluded), subjected to 24 genera of 21 families of eight orders of five classes of four phyla, which were Cnidaria, Arthropoda, Chaetognaths, and Urochordata. There were nine species of medusa, 16 species of Copepods, two species Cladocera, one species Amphipoda, one species Decapod, one species Chaetognatha, one species Tunicate, and 17 species of larvae and juveniles (Ocean & Fishery Administration of Qingdao, 2002–2015). The average density of zooplankton was 393 ind.·m^{-3}, the average biomass was 457 mg m^{-3}, and the average biodiversity index was 2.78. There were 75 species of macrobenthos obtained, belonging to 68 genera of 52 families of 30 orders of 12 classes of seven phyla. The dominant genus was polychetes with 30 species. There were 22 species of crustacean, 13 species of mollusk, six species of echinoderm, two species of Cnidaria, one species of Nemertea, and one species of Brachiopoda. The average density of benthos was 10^5 ind.·m^{-2}, the average biomass was 47.0 g m^{-2}, and the average biodiversity index was 2.01 (Ocean & Fishery Administration of Qingdao, 2002–2015).

2 Environmental Responses to Human Interventions

2.1 Coastline Change

Based on remote sensing images, historical charts, and water depth observations, JZB's coastline and sea area change have been investigated. During the period from 1863 to 1935, the coastline changed little, and the sea area only decreased 15.65 km^2 in 72 years, from 567.95 km^2 to 552.3 km^2, which indicates a natural evolution-dominated stage (Zhou et al., 2010).

Thereafter, JZB was dramatically influenced by human activities (Zhou et al., 2010; Ma et al., 2014). Large areas of natural intertidal flat were replaced by salt ponds, aquaculture, and reclamation. Hongdao Island and Huangdao Island were connected to the mainland in 1966 and 1986, respectively. The coastline tended to be straighter in 2013 (Fig. 1A). During the period from 1935 to 1986, natural coastline and intertidal area decreased sharply by 137.4 km and 136.8 km^2 (Fig. 1B), about 70% and 50% of that in 1935, respectively. From 1987 to 2002, the rate of sea area change reached a maximum value, induced by a total of 20 reclamations in JZB, with a total area of 27.67 km^2 (Ma et al., 2014). After 2002, land reclamation in the inner bay gradually stopped, but rapid reduction still occurred in the outer bay due to the construction of numerous ports and docks (Yuan et al., 2016).

In total, during the period from 1935 to 2013, the artificial coastline increased by 138.6 km, which was 7.5 times more than that in 1935. Furthermore, sea area decreased by 213 km^2 in 78 years, which indicated that reclamation and harbor construction reduced more than one third of the embayment area (Wang et al., 2014).

2.2 Hydrodynamics

Numerical modeling is often conducted to study anthropogenic influences on hydrodynamics. Based on an unstructured grid, Finite Volume Community Ocean Model (FVCOM; Chen et al. 2003), the hydrodynamic environment between 1935 and 2008 was compared. The 30-day averaged Eulerian current filed illustrated that the circulation system in JZB had changed a lot from 1935 to 2008. The multi-vortex pattern of the residual current in JZB remained the same, but the maximum velocity speed had reduced from 72.1 cm s^{-1} to 49.5 cm s^{-1}. The anticyclonic eddy at the bay mouth had decreased dramatically, which indicated weakened capability to transport sediments or nutrients from the inner bay to the outer bay. For example, the water circulation was rather robust in the northeastern part, where the worst water quality in JZB is found now (Yuan et al., 2016).

Using the same numerical model, Gao et al. (2014) found that the M$_2$ tidal-energy flux across the entrances to the inner bay had been reduced by more than 50%, also indicating a decrease of water exchange capability. Furthermore, the half exchange time had been extended by 10.3% as the tidal prism had been reduced by 31.5% from 1935 to 2008 (Fig. 1B; Chen and Chen, 2012). The decrease of water exchange capability had increased the average residence time over the past decades, particularly after the 1980s (Shi et al., 2011). Such changes have caused the accumulation of gradually increased terrigenous pollutants along the coast and the deterioration of water quality and ecosystem in the bay (Yuan et al., 2016).

2.3 Sediment Transport

A three-dimensional, high-resolution FVCOM tidal model coupled with the University of New South Wales Sediment model (UNSW-Sed) was applied to study the transport change of suspended sediments between JZB and its adjacent sea areas with respect to land reclamation over the period of 1935 to 2008 (Gao et al., 2017).

The tidal flats are a primary source for sediment resuspension, leading to turbidity maxima nearshore. With the process of land reclamation over the past seven decades, the suspended sediment concentration (SSC) was weakened due to reduced tidal currents. In 1935, the net movement of suspended sediments was from JZB to the adjacent seas (erosion for JZB), primarily caused by horizontal advection associated with a horizontal gradient in the SSC. This seaward transport (erosion for JZB) had gradually declined from 1935 to 2008 (Gao et al., 2017). If land reclamation on a large scale continued in the future, the net transport between JZB and the adjacent seas would turn landward and erosion for JZB would switch to siltation (Gao et al., 2017). Taking the bay mouth (which is important for ship route with deep water) as an example, the situation had been transferred from erosion before the 1990s to siltation after the 1990s (Shi, 2010).

2.4 Water Quality

Since the 1960s, especially since the beginning of the 1980s with the rapid population growth and economic development in Qingdao, water quality of JZB has been deteriorating. Annual mean DIN concentration in seawater of JZB rose about 13 times from the 1960s to the 2010s, and now exceeds the second level of CNSQS ($0.3\,\text{mg}\,\text{L}^{-1}$) (Fig. 4) (Zhang et al., 2017). Meanwhile,

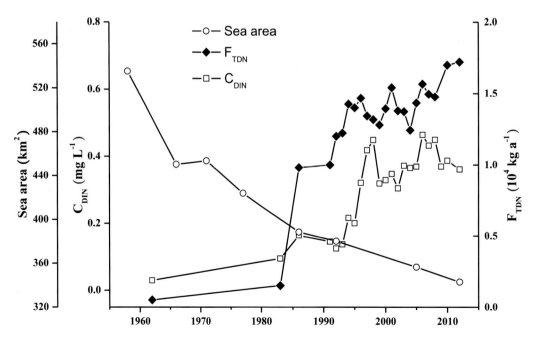

Fig. 4 Long-term variation of flux of total dissolved nitrogen (F_{TDN}) and annual average concentration of dissolved inorganic nitrogen (C_{DIN}) with sea area change during the last six decades.

active phosphate concentration was increasing until the mid-1990s, and exceeded the first level of CNSQS in the 2010s ($0.015 \, \text{mg} \, \text{L}^{-1}$) (Liang et al., 2015), while COD concentration increased slightly and kept below the first level of CNSQS ($2.0 \, \text{mg} \, \text{L}^{-1}$) (Liang et al., 2015).

On the basis of land-sea interaction (Borja et al., 2006), seawater quality is the integrated responses of both anthropogenic pressure indicators (APIs) and coastal carrying indicators (CCIs) (Crain et al., 2008; Borja et al., 2016). Taking DIN, which is the dominant pollutant in JZB, as an example, variation of water quality and its relationships with APIs and CCIs were identified using long-term time series data (1962–2012). High-intensity land-based source TDN discharge and land reclamation were the important factors of long-term variation of DIN concentration (Zhang et al., 2017). More than 60% of land-based pollutants from Qingdao were discharged into JZB, which contributed greatly to water quality deterioration in JZB (Wang et al., 2006). From the 1960s to the 2010s, land-based wastewater discharge flux and TDN flux had increased about 300% and 1000%, respectively (Fig. 4) (Wang et al., 2006; Qian et al., 2009; Dong et al., 2010; Liang et al., 2015). During the same period, intensive industry and population increase were aggregated in the coastal zone around JZB, resulting in increased municipal and industrial discharges as well as excessive agricultural fertilizer loss, which were responsible for the dramatic growth of the pollutant load entering JZB (Zhang, 2007; Liang et al., 2015). On the other hand, extensive reclamation around JZB was executed to meet the growing demands for land development, leading to decline in the tidal prism and weakness in water exchangeability (Fig. 4) (Shi et al., 2011; Gao et al., 2014; Yuan et al., 2016). The reduction of self-purification capability resulted in continuous water quality degradation and ecosystem deterioration in JZB. Moreover, the water quality simulation indicated there is a synergistic interaction of terrigenous pollutant discharge and land reclamation regarding water quality of JZB (Zhang et al., 2017). To effectively improve the water quality of JZB, it is necessary to both control terrigenous pollutant load and implement scientific plans for coastline utilization.

2.5 Biology

2.5.1 Succession of Species, Biomass, and Biodiversity of Halobios

During the last six decades, the biological system of JZB, including plankton, nekton, and benthos habitats, also varied greatly under the multiple pressures of natural changes and human activities. The total abundance of phytoplankton increased evidently from the 1980s, affected by the increase in nutrient concentration and the predation pressure of maricultured shellfish in JZB (Sun et al., 2011a). Meanwhile, the phytoplankton community of JZB had changed in response to the impacts of global warming and nutrient compositions, which were conspicuously reflected by the dramatic increase of the small chained diatom species, the nitrophilous species, the warm water species, and the dinoflagellate species (Wu et al., 2005; Sun et al., 2011b). The biomass of zooplankton showed a dramatically increasing trend, too. The dominant zooplankton species is Copepoda (Sun et al., 2011c). The zooplankton biodiversity after 2000 was 30% higher than before 2000 (Sun et al., 2008, 2011c). The species diversity of macrobenthic fauna also changed from the 1980s, which was highest in 1981, but then declined and reached its lowest point in 1991, and almost recovered in the 2000s (Yu et al., 2006; Wang et al., 2011). The dominant microbenthic fauna in JZB are *Ruditapes philippinarum*, *Temnopleurus toreumaticus* and *Protankyra bidentata*. Among them, the population of *R. philippinarum* has been dominant by means of aquaculture in recent years (Yu et al., 2006). A total of 113 species of marine fish was collected and identified from JZB area during the exploratory trawling carried out in the years 1980–1984. They belonged to 12 orders, 52 families and 90 genera. The resident included a number of economic species, which could be fished all the year round, and so was of great importance to local fishery production. However, only 58 species were recorded in the years 2003–2004 (Zeng et al., 2004). Most of them were low-value fishes (Zeng et al., 2004). These results indicated that the production function, regulatory function, and biological diversity of JZB have declined during the last three decades.

2.5.2 Ecological Disasters

Human activities have led to the aggravation of eutrophication in coastal waters of JZB, and trigger various ecological disasters, such as tide microalgal or green macroalgal blooms. The eutrophication index (EI), used to show the propensity for undesirable algal blooms, shows that JZB reached a threshold in the early 1980s, became serious from the early-1990s and is worse now (Dai et al., 2007; Qian et al., 2009). Algal blooms were not recorded in JZB before 1990; however, they occurred almost each year after 1991. From the mid-1990s to the early-2000s, both the frequency and area of algal bloom occurrence increased year by year (Wu et al., 2005). Moreover, the large-scale green bloom

(*Enteromorpha prolifera*) occurred each year in JZB and Qingdao coastal from 2007 (Ocean & Fishery Administration of Qingdao, 2002–2015), which indicated an aggravating trend of algal or green bloom disaster due to eutrophication.

3 Environmental Management

3.1 Case Study: Cross-Sea Bridge

To reduce urban traffic and promote economic development, the construction of the JZB Cross-sea Bridge began in 2007, and the bridge opened to traffic in 2011 (Fig. 1A). The total length of the JZB Bridge is 25.171 km, with 952 piers bored into the sea area. For the two largest waterways, the Dagu River channel and the Cangkou channel (Fig. 1A), the piers were built 260 m away in distance, which is much larger than the other piers of 50–60 m in space.

A high-resolution numerical model was established by Zhao (2016) for the cases before and after the bridge construction, based on the MIKE 21 Flow Modules by Danish Hydraulic Institute (DHI). Due to the bridge construction, the difference of tidal elevation between the waters inside and outside of the bay increased. However, both tidal current and residual current decreased significantly at the bay entrance. During the spring tide, tidal prism was decreased by 1.7%, induced by the pier blockage effect, which equals 140 times more than the pier's bulk in the tidal range (Li et al., 2014a,b). The seabed sediment became coarser on the north side of the bridge in western JZB, and became finer on the south side. Coarser sediment belts can be found across the channel bridge (Zhao et al., 2015) in both sides. The capacity of seawater exchange in JZB has been greatly affected by the construction of the cross-sea bridge. Especially in the Hongdao Island coastal area, the 30-day water exchange rate has decreased by 20%, which may affect the environment significantly by decreasing pollutant diffusion, even increasing the amount of floe-ice north of the bridge in winter (Zhao, 2016).

Actually, the environment impact was simulated before the bridge construction in 1993 (Sun et al., 1994a,b,c); however, negligible change was predicted by the crude numerical model, which was newly developed at that stage. Li et al. (2014a,b) and Zhao (2016) used high-resolution models to show that the accumulated blockage effect of the whole bridge could induce hydrodynamic change obviously, especially at the bay entrance, though there is only tiny impact due to a single pier. Thus, to slow down environmental deterioration, land reclamation and sewage discharge should be reduced, especially in the northern bay.

3.2 Case Study: Hongdao Channel

Most reclamation took place at the head of JZB between 1935 and 1966, with Hongdao Island being connected to the mainland. To evaluate the effect of the Hongdao Channel on the environmental change, a three-dimensional barotropic model based on the FVCOM was set up (Gao et al., 2014). According to the numerical experiments, the effects of tidal-flat and Hongdao Island reclamation between 1935 and 1966 were identified.

The M_2-M_4 tidal-duration asymmetry in JZB was significantly more affected by the reclamation of tidal flats than by the connection of Hongdao Island to the mainland (Gao et al., 2014). Furthermore, the tidal asymmetry change caused by the loss of tidal flats and by Hongdao Island-mainland connection was not simply a linear summation of individual changes caused by the two events (Gao et al., 2014). A debate has taken place in recent years on whether an artificial channel should be engineered across the Hongdao Peninsula to improve circulation and overall water quality in northeastern JZB. This case study showed that restoring the Hongdao Chanel will not significantly change the hydrodynamic processes of the bay, thus it will not be a good investment to improve the environment.

3.3 Politics and Laws

As part of China's Blue Economic Strategy, JZB is experiencing accelerated socioeconomic development (Liang et al., 2015); as a result, JZB is under severe stress from intensive human activities. In recent decades, the Qingdao Municipal Government has always played an active role in protecting the marine environment and ecological values of JZB. Various legal and political measures including laws, regulations, and guidelines at national, provincial, and local levels have been implemented on JZB.

For example, Qingdao was among the first regions in China to legally intervene in coastal zone management, and several regulations have been established since 1995, such as the *Nearshore Marine Environmental Protection of Qingdao*, the *Regulation on Coastal Zone Planning of Qingdao*, the *Coastal Functional Zoning of JZB and Its Vicinity*, and the *Regulation on Marine Environment Protection of Qingdao* (Liang et al., 2015; Yuan et al., 2016). In particular, "Regulation on the Protection of Jiaozhou Bay," which is considered the most stringent protection regulation in history, has been in place since September 1st, 2014. A committee has also been established to coordinate all JZB related projects (Yuan et al., 2016). Additionally, in the year of 2009, the Jiaozhou Bay Coastal Wetland Marine Special Reserve at provincial level was established, which is the first marine special reserve in Qingdao (Qiao, 2009).

In spite of the political focus, extensive effort, and achievement in the protection of marine ecological environment and management of resource exploitation in JZB (Ocean & Fishery Administration of Qingdao, 2002–2015), there are still many problems; the ecological trend has not been reversed, and the management is far from a success. Knowledge gaps still exist in scientists' understanding of the changing mechanisms of JZB's environment, and their ability to quantify the impact of human activities on the biogeochemical processes (Yuan et al., 2016). Therefore, further research and integrated management solutions are needed for such an area with complicated and complex systems. A few recommendations are provided as follows for JZB's sustainable coastal development in the future.

4   Conclusions

Economic development of the coastal city of Qingdao, China, largely depends on the resources supplied by JZB, which has large area wetlands, abundant halobios, deep waterways, and so on. However, the bay environment obviously changed since 1935 due to intensive human activities, including land reclamation, salt ponds, harbors, and cross-bay bridge construction. The environmental change and management of JZB received wide scientific and management attention through field observations, satellite observations, and numerical model simulations.

Until 2013, JZB's sea area decreased by $213\,km^2$, with its straighter coastline; more than one third of the embayment area was occupied by reclamation and harbor construction. Furthermore, the half water exchange time had been extended by 10.3% as the tidal prism decreased by 31.5% from 1935 to 2008, indicating a decrease of water exchange capability. Such changes caused the accumulation of gradually increased terrigenous pollutants along the coast and the deterioration of water quality and ecosystem in the bay. Furthermore, as an important ship route, the bay mouth has been transformed from erosion to siltation after the 1990s due to the bay's reduced hydrodynamics.

To effectively improve the water quality of JZB, it is necessary to both control terrigenous pollutant discharge and implement scientific plans for coastline utilization. In recent decades, the central government and Qingdao municipal government have recognized these serious problems. Various legal and political measures including laws, regulations and guidelines at national, provincial, and local levels have been implemented on JZB to protect its marine environment and ecological values.

Nevertheless, to avoid the degeneration of JZB's marine environment and to restore its ecological values, lessons should be learned from history, and recommendations are given. First, a commonly applied management model for JZB and an integrated indicator framework incorporating biophysical, economic, and social indicators should be built to better assist in scientific research and management. Real-time and long-term monitoring, prediction, and databases integrating geospatial, environmental, ecological, socio-economic, policy, and management information should be established. The capability to convert all the information into knowledge is urgently needed (Liang et al., 2015). Second, extensive/systematic research on human activities' environmental impact on JZB should be continuously conducted. As suggested by previous studies (e.g., Yan, 2003; Sun, 2008; Wang, 2009, 2013; Yu, 2010). Third, Integrated Coastal Zone Management (ICZM) should be implemented and strengthened with the support of an interdisciplinary problem-solving mechanism to promote ongoing development of government, information and knowledge sharing, stakeholder participation, and public awareness regarding JZB. Finally, international collaborations are also needed to share understanding of similar research and management cases all over the world for common improvements.

Now JZB has become part of China's Blue Economic Strategy. The insights and recommendations from JZB can be highly transferable to other bays/harbors in China as well as those in developing countries.

References

Bai, W.M., 2005. Research on engineering Geond ons from JZB vironment and the Jiaozhou Gulf. Ocean University of China, Qingdao (in Chinese with English abstract).

Borja, Á., Elliott, M., Andersen, J.H., Berg, T., Carstensen, J., Halpern, B.S., 2016. Overview of integrative assessment of marine systems: the ecosystem approach in practice. Front. Mar. Sci. 3, 20.

Borja, Á., Galparsoro, I., Solaun, O., Muxika, I., Tello, E.M., Uriarte, A., 2006. The European water framework directive and the DPSIR, a methodological approach to assess the risk of failing to achieve good ecological status. Estuar. Coast. Shelf Sci. 66, 84–96.

Chen, C., Liu, H., Beardsley, R.C., 2003. An unstructured grid, finite volume, three-dimensional, primitive equations ocean model: application to coastal ocean and estuaries. J. Atmos. Ocean. Technol. 20, 159–186.

Chen, J., Chen, X., 2012. Numerical simulation of the hydrodynamic evolution of the Jiaozhou Bay in the last 70 years. Acta Oceanol. Sin. 34, 30–41 (in Chinese with English abstract).

Crain, C.M., Kroeker, K., Halpern, B.S., 2008. Interactive and cumulative effects of multiple human stressors in marine systems. Ecol. Lett. 11, 1304.

Dai, J.C., Song, J.M., Li, G.X., Yuan, H.M., Li, N., Zheng, G.X., 2007. Environmental changes reflected by sedimentary geochemistry in recent hundred years of Jiaozhou Bay, North China. Environ. Pollut. 145, 656.

Dong, Z., Lou, A., Cui, W., 2010. Assessment of eutrophication of Jiaozhou Bay. Mar. Sci. 34, 35–39 (in Chinese with English abstract).

Editorial Committee for China's Harbours and Embayments, 1993. China's Harbours and Embayments (Part 4): Southern Shandong Peninsula and Jiangsu Province. China Ocean Press, Beijing, pp. 157–258 (in Chinese).

Fu, M.Z., Li, Z.Y., Gao, H.W., 2007. Distribution characteristics of nonylphenol in Jiaozhou Bay of Qingdao and its adjacent rivers. Chemosphere 69, 1009–1016.

Gao, G.D., Wang, X.H., Bao, X.W., 2014. Land reclamation and its impact on tidal dynamics in Jiaozhou Bay, Qingdao, China. Estuar. Coast. Shelf Sci. 151, 285–294.

Gao, G.D., Wang, X.H., Bao, X.W., Song, D., Lin, X.P., Qiao, L.L., 2017. The impacts of land reclamation on suspendedamics in Jiaozhou Bayngdao and its, Qingdao, China. Estuar. Coast. Shelf Sci. (Accepted).

Gao, S., Wang, Y.P., 2002. Characteristics of sedimentary environment and tidal inlet evolution of Jiaozhou Bay. Adv. Mar. Sci. 20, 52–54 (in Chinese with English abstract).

Han, H.Y., Li, K.Q., Wang, X.L., Shi, X.Y., Qiao, X.D., Liu, J., 2011. Environmental capacity of nitrogen and phosphorus pollutions in Jiaozhou Bay, China: modeling and assessing. Mar. Pollut. Bull. 63, 262–266.

Jiang, D., Wang, X., 2013. Variation of runoff volume in the Dagu River basin in the Jiaodong Peninsula. Arid Zone Res. 30, 965–972 (in Chinese with English abstract).

Jin, B., Gong, L., Song, J., 2010. Heavy mineral analysis in the sediment originated from the Daguhe River and its environmental significance. Mar. Sci. 34, 71–76 (in Chinese with English abstract).

Li, G.X., Liu, Y., Shi, J.H., Dong, H.P., Ma, Y.Y., Li, P., Luan, G.Z., 2014a. Geology and Environment of the Jiaozhou Bay. Ocean Press, Beijing (in Chinese).

Li, P., Li, G.X., Qiao, L.L., Chen, X.E., Shi, J.H., Gao, F., Wang, N., Yue, S.H., 2014b. Modeling the tidal dynamic changes induced by the bridge in Jiaozhou Bay, Qingdao, China. Cont. Shelf Res. 84, 43–53.

Li, S.W., Wang, Y.J., Zhang, Q.N., Xu, X.S., 1986. The geomorphological development of the Jiaozhou Bay. Mar. Sci. Bull. 5, 10–18 (in Chinese with English abstract).

Liang, S.K., Pearson, S., Wu, W., Ma, Y.J., Qiao, L.L., Wang, X.H., Li, J.M., Wang, X.L., 2015. Integrated coastal zone management of Jiaozhou Bay, China: review of research and management with recommendations to sustain the Bay's functions. Ocean Coast Manag. 116, 470–477.

Ma, L.J., Yang, X.G., Qi, Y.L., Liu, Y.X., Zhang, J.Z., 2014. Oceanicarea change and contributing factor of Jiaozhou Bay. Sci. Geogr. Sin. 3, 365–369 (in Chinese with English abstract).

Ocean & Fishery Administration of Qingdao, 2002–2015. Report on Marine Environmental Quality of Qingdao. http://ocean.qingdao.gov.cn. (in Chinese).

Qian, G.D., Han, H.Y., Liu, J., Liang, S.K., Shi, X.Y., Wang, X.L., 2009. Spatiotemporal changes of mainchemical pollutants for the last thirty years in the Jiaozhou Bay. J. Ocean Univ. 39, 781–788 (in Chinese with English abstract).

Qiao, X., 2009. The Study on Jiaozhou Bay Discharge Regions and Accurate Calculationsof their Allocated Capacities of Major Pollutants. Ocean University of China, Qingdao.

Sheng, M., Cui, J., Shi, Q., Li, L., Geng, Y., 2014. Analysisi of sediment discharge characteristics of reivers in Jiaozhou Bay, Qingdao, China. J. China Hydrol. 34, 92–96 (in Chinese with English abstract).

Shi, J., Li, G., Wang, P., 2011. Anthropogenic influences on the tidal prism and water exchanges in Jiaozhou Bay, Qingdao, China. J. Coast. Res. 27, 57–72.

Sun, L., 2008. Coastal Ecosystem Health Assessment and Prediction Research of Jiaozhou Bay. Ocean University of China, Qingdao.

Sun, S., Li, C.L., Zhang, G.T., Sun, X.X., Yang, B., 2011c. Long-term changes in the zooplankton community in the Jiaozhou Bay. Chin. J. Oceanol. Limnol. 42, 625–631 (in Chinese with English abstract).

Sun, S., Zhou, K., Yang, B., Zhang, Y.S., Ji, P., 2008. Ecology of zooplankton in the Jiaozhou Bay I. Species composition. Chin. J. Oceanol. Limnol. 39, 1–7 (in Chinese with English abstract).

Sun, X., Sun, S., Wu, Y., Zhang, Y., Zheng, S., 2011b. Long-term changes of phytoplankton community structure in the Jiaozhou Bay. Chin. J. Oceanol. Limnol. 42, 639–646 (in Chinese with English abstract).

Sun, X., Sun, S., Zhang, Y., Zhang, F., 2011a. Long-term changes of chlorophyll-a concentration and primary productivity in the Jiaozhou Bay. Oceanologia et Limnologia Sinica 42, 654–661 (in Chinese with English abstract).

Sun, Y.L., Sun, C.Q., Wang, X.C., Tian, H., Chen, S.J., 1994a. Forecast of impact of Qingdao Bay bridge on tide, tidal current and residual current of Jiaozhou Bay. Tidal current of Jiaozhou Bay and adjacent sea area. J. Ocean Univ. Qingdao, 105–119 (in Chinese with English abstract).

Sun, Y.L., Tian, H., Zheng, L.Y., Chen, S.J., 1994b. Forecast of impact of Qingdao Bay bridge on tide, tidal current and residual current of Jiaozhou Bay II. Study of forecast method. J. Ocean Univ. Qingdao, 120–125 (in Chinese with English abstract).

Sun, Y.L., Wang, X.C., Sun, C.Q., Chen, S.J., 1994c. Forecast of impact of Qingdao Bay bridge on tide, tidal current and residual current of Jiaozhou Bay III. Forecast method and forecast results. J. Ocean Univ. Qingdao, 126–133 (in Chinese with English abstract).

Wang, C., 2009. Study of the Pattern for Integrated Coastal Zone Management Based on Ecosystem Approach: A Case Study of Jiaozhou Bay. Ocean University of China, Qingdao.

Wang, C.Y., Liang, S.K., Li, Y.B., Li, K.Q., Wang, X.L., 2015. The spatial distribution of dissolved and particulate heavy metals and their response to land-based inputs and tides in a semi-enclosed industrial embayment: Jiaozhou Bay, China. Environ. Sci. Pollut. Res. 22, 10480–10495.

Wang, H.F., Li, X.Z., Wang, J.B., 2011. Macrobenthic composition and its changes in the Jiaozhou Bay during 2000-2009. Oceanologia et Limnologia Sinica 42, 738–752 (in Chinese with English abstract).

Wang, X., 2013. In: A brief introduction to the "Recommendation on Prompting Marine Ecological Civilisation Construction in the Next Five Years in Qingdao" the President's Council of Qingdao's People's Political Consultative Conference Proposal to Qingdao Government. Annual Meeting Sinonual Meeting Sinorence Proposal to Qingdao Government, Annua. Ocean University of China, Qingdao.

Wang, X.L., Li, K.Q., Shi, X.Y., 2006. The Marine Environmental Capacity of Pollutants in Jiaozhou Bay. Science Press, Beijing, pp. 3–20 (in Chinese).

Wang, Y.P., Gao, S., Jia, J.J., 2000. Sediment distribution and transport patterns in Jiaozhou Bay and adjoining areas. Acta Geograph. Sin. 55, 449–458 (in Chinese with English abstract).

Wang, Y.P., Gao, S., Jia, J.J., Liu, Y.L., Gao, J.H., 2014. Remarked morphological change in a large tidal inlet with low sediment-supply. Cont. Shelf Res. 90, 79–95.

Weng, X., Zhu, L., Wang, Y., 1992. Physical oceanography. In: Ecology and Living Resources of Jiaozhou Bay. Science Press, Beijing, pp. 20–72 (in Chinese).

Wu, Y.L., Sun, S., Zhang, Y.S., 2005. Long-term change of environment and its influence on phytoplankton community structure in Jiaozhou Bay. Chin. J. Oceanol. Limnol. 36, 487–498.

Xu, X.Q., Yang, H.H., Li, Q.L., Yang, B.J., Wang, X.R., Lee, F.S.C., 2007. Residues of organochlorine pesticides in near shore waters of Laizhou Bay and Jiaozhou Bay, Shangdong Peninsula, China. Chemosphere 68, 126–139.

Yan, J., 2003. Research on Integrated Coastal Zone Management in Jiaozhou Bay. Ocean University of China, Qingdao.

Yan, X.X., Wu, M.Y., Liu, G.T., 2000. Analysis of morphological characteristics and seabed processes of Jiaozhouwan Bay. J. Waterw. Harb. 4, 23–29 (in Chinese with English abstract).

Yu, H., Li, X., Li, B., Wang, J., Wang, H., 2006. The biodiversity of macrobenthos from Jiaozhou Bay. Atca Ecological Sina 26, 416–422 (in Chinese with English abstract).

Yu, Y., 2010. Integrated Coastal Area and River Basin Management in Jiaozhou Bay. Ocean University of China, Qingdao.

Yuan, Y., Song, D., Wu, W., Liang, S., Wang, Y., Ren, Z., 2016. The impact of anthropogenic activities on marine environment in Jiaozhou Bay, Qingdao, China: a review and a case study. Regional Stud. Mar. Sci. 8, 287–296.

Zeng, X., Piao, C., Jiang, W., Liu, Q., 2004. Biodiversity investigation in Jiaozhou Bay and neighbouring waters. J. Ocean Univ. China 34, 977–982 (in Chinese with English abstract).

Zhang, J., 2007. Watersheds nutrient loss and eutrophication of the marine recipients: a case study of the Jiaozhou Bay, China. Water Air Soil Pollut. Focus 7, 583–592.

Zhang, M.H., 2000. Distributions and seasonal variations of suspended matter in Jiaozhou Bay seawater. Studia Marina Sinica 42, 49–54 (in Chinese with English abstract).

Zhang, P., Su, Y., Liang, S.K., Li, K.Q., Li, Y.B., Wang, X.L., 2017. Assessment of long-term water quality variation affected by high-intensity land-based inputs and land reclamation in Jiaozhou Bay, China. Ecol. Indic. 75, 210–219.

Zhang, P., Wang, C.Y., Liang, S.K., Wang, X.Y., Wang, X.L., 2016. Distribution characteristics and ecological risk assessment of Nonylphenol in the Jiaozhou Bay in Qingdao, China. Fresenius Environ. Bull. 25, 5432–5439.

Zhang, Z.H., 2009. Utilization Status and Assessment of Coastal Zone in Jiaozhou Bay. Ocean University of China (in Chinese with English abstract).

Zhao, K., Qiao, L.L., Shi, J.H., He, S.F., Li, G.X., Yin, P., 2015. Evolution of sedimentary dynamic environment in the western Jiaozhou Bay, Qingdao, China in the last 30 years. Estuar. Coast. Shelf Sci. 163, 244–253.

Zhao, K., 2016. The Impact on the Changes of Sedimentary Dynamic Environment Introduced by Jiaozhou Bay Bridge. Ocean University of China (in Chinese with English abstract).

Zhou, C.Y., Li, G.X., Shi, J.H., 2010. Coastline change of Jiaozhou Bay over the last 150 years. Period. Ocean Univ. China 40, 99–106 (in Chinese with English abstract).

3

MUDDY COAST OFF JIANGSU, CHINA: PHYSICAL, ECOLOGICAL, AND ANTHROPOGENIC PROCESSES

Jiabi Du*,†, Benwei Shi‡, Jiasheng Li‡, Ya Ping Wang‡,§
Virginia Institute of Marine Science, College of William and Mary, Gloucester Point, VA, USA †Department of Marine Sciences, Texas A&M University, Galveston, TX, USA ‡Ministry of Education Key Laboratory for Coast and Island Development, Nanjing University, Nanjing, China §State Key Laboratory of Estuarine and Coastal Research, East China Normal University, Shanghai, China

CHAPTER OUTLINE
1 **Introduction**   26
2 **Physical Control**   27
   2.1 Hydrodynamics   28
   2.2 Sediment Source   30
   2.3 Sediment Transport   32
3 **Ecological Control**   34
   3.1 Impact of Marsh Introduction on Sedimentation   34
   3.2 Impact of Salt Marsh on Ecosystem   36
4 **Human Intervention**   37
   4.1 Reclamation   37
   4.2 Aquaculture   39
   4.3 Shoreline Protection Engineering   40
   4.4 Damming   40
5 **Weakness of Current Studies on Jiangsu Mudflat**   41
6 **Summary**   44
Acknowledgments   45
References   45
Further Reading   49

Sediment Dynamics of Chinese Muddy Coasts and Estuaries. https://doi.org/10.1016/B978-0-12-811977-8.00003-0

25

1 Introduction

Intertidal mudflat (also known as tidal flat) comprises an important transition zone between marine and terrestrial systems. Intertidal mudflats, especially those with salt marshes, provide some of the most valuable ecosystem services upon which humans and other species depend, such as food supply, recreational and educational resources, nutrient trapping, carbon fixation, and contaminant sequestration (Foster et al., 2013). This natural buffering area also provides the first line of defense against coastal flooding and storm waves (Cooper, 2005). Mudflats are found over the global coast, mainly in the coastal area near the mouth of a large river that supplies a lot of sediments to the ocean. Mudflats in many coastal systems have experienced a landward retreat during the 20th century and are expected to retreat at an accelerated rate in the future, due to coastline erosion under accelerated sea-level rise and global warming (Zhang et al., 2004). Only a few mudflat systems, mostly located near the mouths of large rivers (e.g., Yangtze/Changjiang, Mississippi, Amazon, and Mekong Rivers), are still expanding and advancing seaward, because of continuous and sufficient sediment supply from these rivers (McBride et al., 2007; Anthony et al., 2010; Tamura et al., 2010). The Jiangsu mudflat, located in the southern Yellow Sea, has been advancing seaward for more than one century, which is different from common mudflat systems as there is no large river in this area that can provide sufficient sediments for the continuous expansion. Instead, the continuous seaward advancement of the middle Jiangsu Coast is caused by the unique hydrodynamics in this region and remote sediment supply from the ancient river deltas and offshore sand ridges. Development of the modern Jiangsu mudflat occurred in the past several hundreds of years, but in just a few decades, it has experienced a significant change because of the great influence from human activities (Zhang, 1992).

The Jiangsu Coast starts from the mouth of the Xiuzhen River in the north to the Yangtze River mouth in the south, with a total length of 954 km, of which more than 90% is mudflat (Shen et al., 2006). Even after several intensive reclamations since 1949, the mudflat along the Jiangsu Coast is still dominant, with an area of 880 km^2 (Ren, 1986), which accounts for about 13% of the total mudflat along the entire China Coast (Yang et al., 2002). The width of the mudflat ranges from 3 km along the southern and northern coasts to more than 10 km along the central coast, with an average of 6 km (Fig. 1). Its slope is rather small, ranging from 0.1/1000 to 1.0/1000.

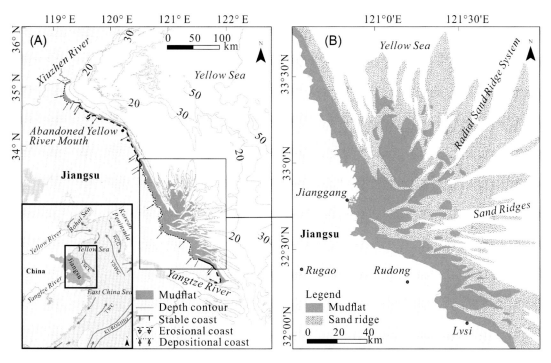

Fig. 1 (A) Map of muddy Jiangsu Coast. (B) Radial sand ridge system off the middle Jiangsu Coast. The ocean currents identified in (A) include the Yellow Sea Coastal Current (YSCC), the Yellow Sea Warm Current (YSWC), the Korea Coastal Current (KCC), and the Taiwan Warm Current (TWC).

The mudflat along the Jiangsu Coast has been influenced by a variety of processes with timescales ranging from days to thousands of years, including tides, monsoons, typhoons, human reclamation, sea-level rise, river channel migration, and fluvial delta development. In this chapter, we will summarize and discuss the important processes related to the development, maintenance, and change of mudflat in the Jiangsu Coast. These processes include the unique physical condition and sediment transport (Section 2), the significant ecological interaction between mudflat and salt marsh (Section 3), and the intensive intervention by human beings (Section 4).

2 Physical Control

Sufficient fine-sediment supply, middle or large tidal range, and sheltered low-energy environment are three of the most essential conditions for the development of mudflat

(Ren et al., 1984; Allen, 2000). These three conditions are well met in the Jiangsu Coast in a manner different from common mudflats found near the mouths of large rivers. Even though there is presently no major river directly discharging into the Jiangsu Coast, the majority of the middle coast is advancing seaward, with a maximum speed of about $200 \, \text{m} \, \text{yr}^{-1}$, whereas the southern and northern coasts are under moderate erosion (Yang et al., 2002). The advancing speed in the middle coast has slowed down in recent decades, while the erosion in the southern and northern coasts has been weakened by numerous coastal protection engineering projects (Yu and Zhang, 1994). The erosion/accretion status of the coast is highly related to the unique tidal and wave dynamics, sediment sources, and sediment transport processes in the region.

2.1 Hydrodynamics

One of most important physical controls on the mudflat of Jiangsu Coast stems from the unique tidal wave systems in this area. Tide in this area is predominantly semi-diurnal, with a tidal range of 3.9–6.7 m (Ren et al., 1984; Ren, 1986). Tidal range, together with surface slope, directly determines the cross-shore width of the tidal flat (Wang and Zhu, 1990). The widest mudflat is found in the central Jiangsu Coast, where the tidal range is the largest. Hydrodynamics along the Jiangsu Coast is controlled by two tidal waves that converge near the Jianggang area. One is the progressive tidal wave from the East China Sea, which propagates northwestward into the southern ridge area. The other is the counterclockwise rotary Kelvin tidal wave in the north (Fig. 2). Convergence of these two tidal waves not only leads to a maximum tidal range in the central coast (i.e., Jianggang), but also causes a converging Stokes drift current due to tidal wave distortion in the shallow water (Xing et al., 2012).

Distortion of the tidal wave over the shallow water generates a strong flood-dominant tidal asymmetry along most of the coast, which is essential for the net landward fine-sediment transport (Dronkers, 1986; Wang et al., 2012a). In-situ measurements in the central coastal waters showed that the ratio between floods' duration and ebbs' duration was about 0.73, the mean flood current speed was about 140% of the ebb current speed, and the mean suspended sediment concentration (SSC) during floods was 125% of that during ebbs (Ren, 1986; Zhang, 1986).

Shelf currents and circulations in the Yellow Sea also play an important role in redistributing the sediment, especially the fine sediment that can be transported over a long distance. The circulation pattern of the Yellow Sea is comprised of two main

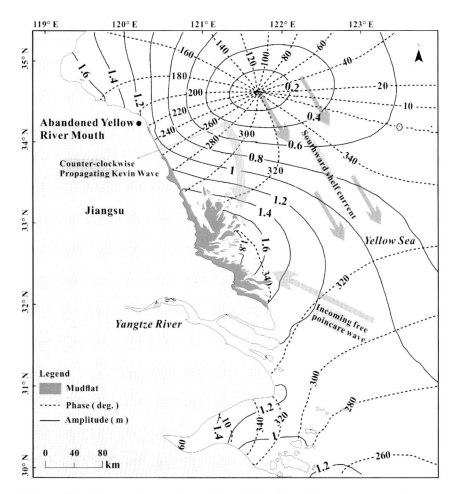

Fig. 2 Co-tidal chart of M₂ tidal constituent in the southern Yellow Sea, with arrows representing the shelf current and two tidal waves' propagation directions. Modified from Xing, F., Wang, Y.P., Wang, H.V., 2012. Tidal hydrodynamics and fine-grained sediment transport on the radial sand ridge system in the southern Yellow Sea. Mar. Geol. 291–294, 192–210.

components, the northward Yellow Sea Warm Current (YSWC) in the east and the southward Yellow Sea Coastal Current (YSCC) in the west (Guan, 1994; Naimie et al., 2001). The circulation pattern varies seasonally, regulated by the seasonal wind (i.e., the typical Asian monsoons), which are characterized with a dominant northeasterly wind in winter and a dominant southwesterly wind in summer. In winter, strong northerly wind drives southward flow at the surface along both Korean and Chinese Coasts. This is compensated by a deep return flow, the YSWC, in the central trough of the Yellow Sea, which can penetrates into the Bohai Sea (Naimie et al., 2001). Influenced by the seasonal wind, the YSCC is strongest during winter, which is confirmed by satellite altimetric data and numerical simulations (Yanagi et al., 1997; Lee and Chao, 2003). The southward YSCC is critical for the sediment supply to the Jiangsu mudflat, as it delivers a lot of sediments eroded near the abandoned Yellow River mouth toward the south, partially supporting the continuous accretion of the mudflat in the middle Jiangsu Coast.

Additionally, the evident seasonality of waves contributes to the seasonal variation of sediment concentration and sediment flux. Waves have been long known to significantly enhance the resuspension of bottom sediment, affecting the sediment transports in both nearshore and offshore waters. Because of the Asian monsoons, wind-induced waves in the South Yellow Sea exhibit significant seasonality. The seasonal wind speeds and directions in summer and winter are distinctly different. Dominant NE-ENE wind in winter has a mean speed of $4.22\,\mathrm{m\,s^{-1}}$ near the sea surface, and dominant SE-SSE wind in summer has a mean speed of $2.76\,\mathrm{m\,s^{-1}}$ (He et al., 2010). Wave height is the smallest in summer and largest in winter under normal conditions. Consequently, SSC is much larger during winter than during summer, with mean SSCs of 375 and $61\,\mathrm{mg\,L^{-1}}$ averaged over the sand ridge area during winter and summer, respectively (Wan and Zhang, 1985). Furthermore, the extent of high SSC is much larger during winter, due to the combination of strong northerly waves and southward shelf current in the northern coast.

Recent measurements showed that wave height is generally lower than 1 m in the sand ridge area (Yang et al., 2014). The small wave height can be attributed to the sheltering from the offshore radial sand ridge system. The radial sand ridge system is comprised of more than 70 sand ridges separated by tidal channels, extending offshore by 30–110 km and covering an area of $22,470\,\mathrm{km^2}$ (Wang et al., 2012b). The radial sand ridge system serves as a huge natural barrier, dissipating a large amount of wave energy. Different wave characteristics between offshore and nearshore favor the suspension and erosion of sediments in the offshore area and deposition of sediment in the nearshore area.

2.2 Sediment Source

Despite moderate erosion in the northern and southern parts of the Jiangsu Coast (Chen, 1990; Zhang et al., 2002), the mudflat is advancing seaward with a maximum speed of about $200\,\mathrm{m\,yr^{-1}}$ near the central coast (Yang et al., 2002). What are the sediment sources that sustain the continuous advancement of the mudflat? Here we discuss sediment sources from four directions, namely, the west (local rivers), north (abandoned Yellow River delta), south (Yangtze River), and east (radial sand ridges).

There is limited sediment input from local rivers along the Jiangsu Coast. Local river input accounted for only 15% of the sediment source for the mudflat in the 1980s (Zhu et al., 1986). Even though the Jiangsu Coast is adjacent to the Yangtze River, which discharges 0.4×10^9 tons of sediments per year into the ocean, sediments from the Yangtze River are largely dispersed eastward into the East China Sea and southward all the way to the Taiwan Strait (Liu et al., 2007; Xu et al., 2009), due to the river plume and the southward shelf current. Only a small portion of Yangtze River sediments could reach south of the sand ridge system during summer (Yuan et al., 2015). Recently, the contributions of the local rivers and Yangtze River were greatly reduced because of the numerous dams built in the upper reaches of these rivers.

The major sediment sources for the Jiangsu mudflat are the sediments eroded in the abandoned Yellow River delta and the offshore radial sand ridge system. The Yellow River discharged to the ocean through the abandoned Yellow River mouth from 1128 to 1855 CE (Wang and Aubrey, 1987; Wang, 2000). During this period, approximately $7–8 \times 10^{12}$ tons of sediment were dumped offshore and the coastline of Jiangsu Province had moved seaward by about 250 km (Zhang, 1986). Although the direct sediment supply from the Yellow River has been cut off since 1855 CE, the large amount of sediment deposited at the submerged abandoned Yellow River delta is still a major sediment source for the modern middle Jiangsu Coast. Additional sediment source is from the ancient Yangtze River delta in the south, particularly near the Lvsi Coast. Sediments from the abandoned Yellow River delta in the north and the Lvsi Coast in the south were estimated to account for about 35% of the total sediment to the mudflat of the middle coast. The other 50% of the sediment came from the offshore submerged sand ridges (Zhu et al., 1986). The contribution percentage differs among different estimations. For example, Wang and Zhu (1990) estimated 0.77×10^9 tons yr^{-1} of sediments were accumulated in the middle Jiangsu Coast, of which 14% came from the abandoned Yellow River delta, 76% came from the offshore sand ridges, and the other 10% came from the modern Yangtze River and the Yellow Sea outside the sand ridge area. Nevertheless, both in-situ measurements and model simulations showed agreement on that the major sediment source for the modern mudflat came from the offshore sand ridges in the east and the abandoned Yellow River delta in the north (Fu and Zhu, 1986; Ren, 1986; Zhang et al., 2013a).

How the radial sand ridges were created in the Holocene is still controversial. Some researchers believed that the ridges were based on the ancient Yangtze River delta (He, 1979; Wang and Zhu, 1990; Wang et al., 2012b), while others hypothesized that the ridges were created by the converging tidal currents during periods when there were abundant sediments (Li et al., 2001; Zhu and Chang, 2001). Seismic images support the former hypothesis because of the existence of ancient river valley beneath the sand ridge area (Wang et al., 2012b). Numerical simulations support the latter hypothesis because of the strong regulation of converging tidal currents on the radial shape morphology, and the fact that the converging tidal currents have existed since the mid-Holocene and are determined by the basin-wide geometry instead of the local radial topography (Zhu and Chang, 2001; Uehara et al., 2002). Sediment in the radial sand ridges is comprised of both Yellow and Yangtze River materials, but the relative prevalence varied in different periods and different regions. Mineralogical analysis of the sediment samples revealed that the sediment of the sand ridges initially came from the Yangtze River and later came from the Yellow River (Li et al., 2001).

The century-long erosion of the abandoned Yellow River delta and sand ridges have caused sediment coarsening and deepening in these two regions. Owing to the reduction of sediment supply from both the Yangtze River delta and abandoned Yellow River delta, the outer part of the sand ridges is likely to suffer from more erosion in the future.

2.3 Sediment Transport

Sediment transport processes in the Yellow Sea have been intensively investigated by geological sampling, hydrodynamic measurement, remote sensing, and numerical modeling since the 1980s. The "settling and scour lag" mechanism, together with the tidal asymmetry, explains both fine- and coarse-sediment transports in the mudflat (McCave, 1970; Zhu et al., 1986). Additionally, modulation of waves is increasingly considered as an important process and has received special attention in advanced numerical models.

The flood-dominant tidal asymmetry over the major part of the Jiangsu Coast causes a net landward sediment flux. The tidal asymmetry has been consistently observed in numerous field measurements. High-frequency measurements carried out at a monitoring station in the central mudflat revealed a landward fine-sediment transport and a seaward coarse-sediment transport, which led to a deposition of finer sediment in the upper mudflat and a deposition of coarser sediment in the lower mudflat

(Wang et al., 2012a). Fine sediment will be transported further toward the upper mudflat during spring tides and the deposited fine sediment in the upper mudflat is usually exposed during neap tides, allowing the compaction of fine sediment in the upper mudflat. The compacted sediment surface in the upper mudflat will not be easily eroded during the following spring tides. Ultimately, the upper mudflat will accrete and the mudflat will advance seaward with time, if sufficient fine sediments are provided.

Wave plays an important role in the initiation of coarse-sediment movement and fine-sediment resuspension and its impact varies significantly in different seasons depending on the wind field (Green and Coco, 2013). Based on field investigations of intertidal sedimentary processes, many researchers have found that "settling and scour lags" were only applicable to intertidal cohesive-sediment transport during periods with weak waves, but not during storms or high-wind events (Shi and Chen, 1996). During winter seasons, strong northerly wind enhanced the sediment suspension in the offshore region, leading to a much larger SSC in winter and a larger spatial extent of high SSC in the radial sand ridge area (Wan and Zhang, 1985). Combined with the converging landward residual current, a large amount of sediment is transported from offshore to the middle mudflat area during winter. In contrast, during summer, the southerly wind with lower speed causes smaller waves and weaker sediment resuspension, resulting in less turbidity and clearer water outside the sand ridge area. The combination of seasonally varying waves and flood-dominant tidal currents causes a net-landward sediment transport over the long term.

It is worthy to note that waves in the offshore area are strongly attenuated by huge sand ridges. In the nearshore area, especially the lower tidal flat, the impact of local waves on sediment resuspension is more profound. Field measurements in the middle Jiangsu Coast indicate that the wave-induced bottom shear stress is essentially important for the erosion/accretion of the mudflat during rough/calm weather condition (Shi et al., 2015, 2016; Xiong et al., 2017).

Additionally, storm events that frequently occur in both summer and winter seasons could enhance the sediment transport flux dramatically. Ren et al. (1983) attributed the silt-sediment deposition in the upper tidal flat in Wanggang to the storm events, during which coarser sediments could be moved to the upper tidal flat. The cross-shore profile of the mudflat can be destroyed and severely reshaped during storm events, as the erosion rate at the lower mudflat during storms could be 10–20 times the erosion rate under a normal condition.

3 Ecological Control

Mudflats of the Jiangsu Coast has supported rich wildlife habitats. Abundant natural living resources in the mudflat and adjacent coastal waters have supported the local economy for centuries. The ecosystem of the Jiangsu Coast is characterized by widespread salt marsh. However, the ecosystem has been significantly changed during the past half century because of the changing climate, natural and anthropogenic alterations.

Halophytic vegetation develops to form salt marsh on the upper intertidal mudflat (Foster et al., 2013). Salt marshes sequester fine sediments along the open coast and on the margins of tidal embayment and estuaries, serving as a natural barrier by dampening storm waves and slowing flows pushing inland (Allen, 2000). In contrast to the widely seen landward retreat of salt marsh over the globe, salt marsh in the Jiangsu Coast is advancing seaward, because of the continuing seaward expansion of the mudflat. Nevertheless, the salt marsh in the Jiangsu mudflat has experienced a rapid reduction in recent years due to extensive reclamation activities.

3.1 Impact of Marsh Introduction on Sedimentation

Different species of grasses dominate in different tidal zones, with *Suaeda salsa* and *Suaeda glauca* in the upper part of the tidal flat, and *Spartina alterniflora* (*S. alterniflora*) in the lower part of the tidal flat between mean-water level and mean high-water level (Fig. 3). There is a strong interaction between sedimentation and salt-marsh expansion on the mudflat of the Jiangsu Coast. The most profound sediment accumulation occurs on the lower part of the tidal flat, particularly on the areas covered by *S. alterniflora*.

Spartina anglica (*S. anglica*, native to the UK) and *S. alterniflora* (native to the Atlantic and Gulf Coasts of North America) were introduced to the Jiangsu Coast in 1963 and 1979, respectively, to mitigate coastline erosion, enhance sediment accumulation, and improve soil quality (Zhong and Zhuo, 1985; Zhuo and Xu, 1985; Fig. 4A). Naturally, after several decades of accretion of the mudflat, sediment accumulation rate of the mudflat is supposed to slow down and the cross-shore profile approaches an equilibrium status, considering the reduced sediment input from the abandoned Yellow River delta and radial sand ridges. However, with the help of *Spartina*, tidal current is significantly weakened by 20%–60% in the bottom boundary layer, resulting in a lower energy environment, which augments the accumulation of fine sediment (Yang and Chen, 1994).

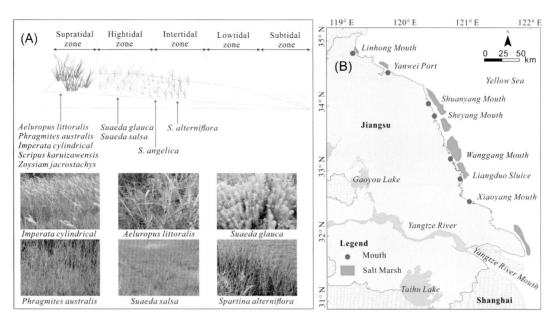

Fig. 3 (A) Salt marsh distribution from the upper tidal flat to lower tidal flat. (B) Salt marsh distribution along the Jiangsu Coast. Modified from (A) Ren, M., Zhang, R., Yang, J., 1984. Sedimentation on tidal mud flat in Wanggang area, Jiangsu Province, China. Mar. Sci. Bull. 3 (1), 40–54; (B) Zhang, R.S., Shen, Y.M., Lu, L.Y., Yan, S.G., Wang, Y.H., Li, J.L., Zhang, Z.L., 2004. Formation of *Spartina alterniflora* salt marshes on the coast of Jiangsu Province, China. Ecol. Eng. 23, 95–105.

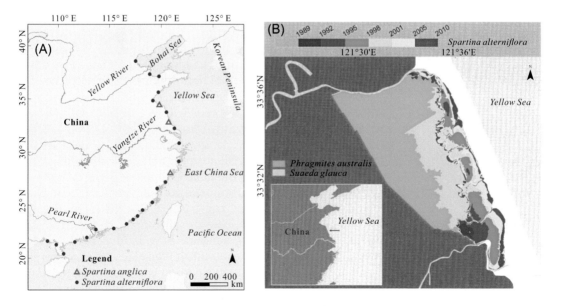

Fig. 4 (A) *S. alterniflora* and *S. anglica* distribution along China coast. (B) Expansion of *S. alterniflora* marsh from 1989 to 2010. Modified from (A) An, S.Q., Gu, B.H., Zhou, C.F., et al., 2007. Spartina invasion in China: implications for invasive species management and future spartina invasion in China. Weed Res. 47, 183–191. (B) After Gao, S., Wang, D., Yang, Y., et al., 2015. Holocene sedimentary systems on a broad continental shelf with abundant river input: process-product relationships. In: Clift, P.D. et al., (Eds.), River-Dominated Shelf Sediments of East Asian Seas.

Measurements of the deposition rate based on ^{210}Pb and ^{137}Cs in Wanggang, located in the middle Jiangsu Coast, showed that the salt marsh had strongly enhanced the deposition rate, with a doubled deposition rate in *S. anglica* marsh and a tripled deposition rate in *S. alterniflora* marsh (Wang et al., 2005). In the Wanggang mudflat, sediment accumulation rate was around $1.4 \, \text{cm} \, \text{yr}^{-1}$ before the introduction of *S. alterniflora*, and the accumulation rate was tripled to $4.2 \, \text{cm} \, \text{yr}^{-1}$ in the years following the introduction (Wang et al., 2005). Impact of salt marsh is even stronger near mouth of rivers. For example, in the mudflat near the mouth of the Yangtze River, sediment accumulation rate was eight times larger over the salt marsh than over the nearby bare flat (Yang and Chen, 1994).

Efficient fine-sediment trapping by *S. alterniflora* affects the distribution of the sediment grain size across the tidal flat. Before the cord grasses were introduced, the bottom sediment across the mudflat had a coarsening trend from the upper to the lower intertidal zone. Field measurements revealed a different pattern of sediment size after the introduction of *S. alterniflora*, with finer sediments on *S. alterniflora* flat in the lower intertidal zone than on *Suaeda salsa* and *S. anglica* flats in the upper intertidal zone (Chen et al., 2005), suggesting an efficient fine-sediment trapping by the *S. alterniflora* marsh.

3.2 Impact of Salt Marsh on Ecosystem

Out of the 954-km Jiangsu Coastline, 410 km is covered by *S. alterniflora*, with a maximum width of 4 km (Zhang et al., 2004). In just 20 years, a total area of 137 km^2 of *S. alterniflora* has covered bare lower tidal flats in the Jiangsu Coast. Despite a decreasing trend in recent decades, the expansion of *S. alterniflora* is rapid (Fig. 4B), with an expansion rate ranging from 10% to 43% annually (Zhang et al., 2004). *S. alterniflora* has now become one of the dominant species on the Jiangsu tidal flat, forming a single-species vegetation system on the large area of the Jiangsu Coast.

At present, there are two opposite perspectives for the introduction of *S. alterniflora* to the Jiangsu Coast. One highlights the positive values from the expansion of salt marsh, including carbon fixation, contaminant degradation, new land generation, and nutrient trapping (e.g., Liu et al., 2007; Wan et al., 2008). It was estimated that 83×10^6 kg of organic carbon were sequestrated annually into the soil pool because of planting *S. alterniflora* in the Jiangsu Coast (Wang et al., 2006). Due to large elongation rates, high leaf area indices, high photosynthetic rates, and long photosynthetic season, *S. alterniflora* enhances the carbon sink capacity of the intertidal ecosystem dramatically and contributes positively to alleviating CO_2 emission into the atmosphere.

On the other hand, due to its strong adaptability and vitality, the *S. alterniflora* community expanded much more rapidly to occupy the lower ecological niche than the native vegetation communities (Zhang et al., 2004), not only colonizing on the adjacent bare flats but also encroaching on the neighboring marsh communities (e.g., *S. anglica* and *Suaeda salsa*). *S. alterniflora* has expanded rapidly from $2.6\,km^2$ in 1985 to $1120\,km^2$ in 2007 along the East Coast of China due to its strong ability to propagate by seeds and rhizome fragments, while *S. anglica* had declined from $360\,km^2$ in 1985 to $0.5\,km^2$ in 2007 (An et al., 2007; Wan et al., 2008). Currently, *S. alterniflora* is the dominant *Spartina* in China and can be found in most of the Chinese mudflats, from Beihai, Guangxi Province, in the south to Tianjin in the north (Fig. 4A). On the basis of its population trend and potential impact on native ecosystems, *S. alterniflora* was officially placed on the list of most harmful invasive alien plants (in total, there are nine species) in China in 2003. *S. alterniflora* invasion in salt marshes has multiple negative effects on the abiotic and biotic properties and on the functioning of the ecosystems, including conversion of mudflats to *Spartina* meadows, loss of shorebirds' foraging habitats, alteration of ecosystem processes (e.g., carbon and nitrogen cyclings), decrease in abundance of native species, degradation of native ecosystems, and considerable economic loss (Wang et al., 2006; Zhou et al., 2008; Zuo and Liu, 2008; Li et al., 2009; Gao et al., 2014).

Nevertheless, it is very unlikely that *S. alterniflora* will be eliminated in the Jiangsu Coast considering the urgent demand for new land, which comes primarily from the *S. alterniflora* marsh flat. In addition, fully destroying *S. alterniflora* may cause even larger consequences to the ecosystem that has already adapted (Xie and Gao, 2009).

4 Human Intervention

Several types of human intervention have affected the sedimentation and ecosystem of the Jiangsu mudflat, including continuous large-scale reclamation since 1949, large areas of aquaculture, and a variety of coastal engineering constructions.

4.1 Reclamation

To meet the increasing demand for new land for agriculture and urban development, hundreds of square kilometers of land are reclaimed each year (Fig. 5). The length of coastline extending

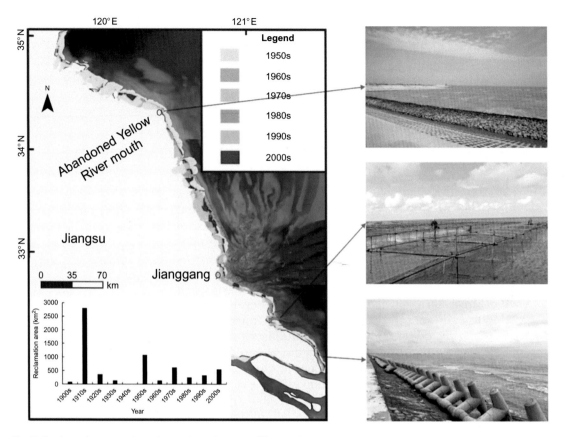

Fig. 5 Reclamation areas from the 1950s to the 2000s. Time series of reclamation area since the 1900s is shown inside the *left panel*; examples of human activities are shown on the *right*, including coastal protection infrastructures and a large area of aquaculture. Modified from Zhang, X., Yan, C., Pan, X.U., Dai, Y., Yan, W., Ding, X., et al., 2013. Historical evolution of tidal flat reclamation in the Jiangsu coastal areas. Acta Geograph. Sin. 68 (11), 1549–1558.

offshore is longer than the China's famous "Great Wall". The new structure protecting the reclaimed land increased 3.4 times in length over 10 years, from 18% to 61% of the total continental coastline, reaching a length of 11,000 km in 2010.

From 1950 to 2004, a total area of 2500 km^2 has been reclaimed (Shen et al., 2006; Fig. 5), which is equivalent to the total reclamation area in the Netherlands over the entire 20th century. The reclamation is ongoing and the reclamation rate is expected to be higher in the coming decades, despite the decreasing seaward-advancing speed (Zhang et al., 2011). The Jiangsu government currently plans to reclaim another 1800 km^2 during 2010–2020. The continuing reclamation largely depends on the sediment accumulation rate in the mudflat area, with maximum reclamation area in the middle coast and minimum reclamation area in the northern and southern coasts. During the tidal flat reclamation, *S. alterniflora* marsh has become the primary target because of the high sediment accumulation rate in the marsh flat. Even though the marsh is still expanding offshore, its expanding speed is far slower than the reclamation rate. Consequently, the tidal flat and salt marsh areas are shrinking, and the mudflat width had decreased from more than 10 km in the 1980s to 2 km presently. Further loss of salt marsh in the future can be forecasted, which may cause a variety of ecological consequences.

The reclamation activities will also affect the tidal dynamics and sediment transport, possibly altering the fine-sediment accumulation rate in the nearshore area (Wang et al., 2012a; Song et al., 2013; Zhu et al., 2016). With a reduced area of tidal flat, the tidal range and phase in the offshore area will be changed, as well as the amphidromic points in the Yellow Sea (Song et al., 2013; Zhu et al., 2016). Bottom sediment became finer on the mid-upper intertidal flat but coarser on the lower intertidal flat following reclamation, in response to a reduction in the tidal currents over the intertidal zone and enhancement of wave action over the lower intertidal flat (Wang et al., 2012a).

4.2 Aquaculture

Global aquaculture contributes 43% of the aquatic animal food for human consumption (Bostock et al., 2010). Aquaculture has a great impact on water quality, nutrient cycling, organic matter accumulation, and sedimentation in coastal zones.

A typical example is the large porphyra cultivation in the nearshore coastal water off Rudong, located in the southern Jiangsu Coast (Fig. 5). The area of porphyra cultivation increased from 56 km^2 in 2003 to 126 km^2 in 2009. Porphyra cultivation will influence the coast morphodynamics and morphological evolution in the near future. During the floods, suspended sediment deposits quickly on the bed in the nearshore and middle parts of the porphyra cultivation zone, resulting in frequent occurrence of accretion events. The main reason lies in the weakened hydrodynamics due to porphyra frames and bamboo rods used in the aquaculture zone. Weakened hydrodynamics prohibit the easy erosion of surface sediment. The influence of porphyra cultivation on bed morphological processes has clear seasonality as porphyra grows from September to April of the following year and the porphyra frames are removed from the bed during the period from April to September every year. Once the porphyra frames are removed from the bed, subsequent erosion of the newly deposited sediments occurs during the following months.

4.3 Shoreline Protection Engineering

To prevent further coastline erosion, large coastal protection engineering infrastructures have been built in both southern and northern parts of the Jiangsu Coast during the 1950s to the 1970s, including seawalls, groynes, and shore parallel offshore breakwaters (Yu and Zhang, 1994; Zhang et al., 2002). With the coastline protection near the abandoned Yellow River mouth, the sediment supply to the central part of the Jiangsu Coast is expected to be reduced significantly and the sustainability of mudflat and salt marsh in the middle Jiangsu Coast is subject to a lot of uncertainties in the future. The reduced retreat rate of coastline near the abandoned Yellow River delta has gradually resulted in accelerated erosion on the lower part of the intertidal zone of the middle coast, even though there is still continuous accretion on the upper part due to the sediment trapping by marsh vegetation. With decades of sediment accumulation in the nearshore area, net sediment flux from the offshore is expected to reduce. In addition, with the accelerated sea-level rise, the amount of sediment needed to sustain the mudflat is increasing. It is therefore unlikely that the advancing speed of the mudflat will remain the same as before.

4.4 Damming

Dams are pervasive features of the world's river systems, and there are more than 400,000 dams over 15 m tall worldwide (Bednarek, 2001). Within the Yangtze River drainage basin, over 48,000 dams have been built, causing significant reduction of sediment discharge and strongly altering the timing of peak freshwater discharge (Gao and Wang, 2008).

As mentioned above, the multiple local rivers delivered about 15% of sediment source to the central Jiangsu Coast in the 1980s. Around 100 sluice gates have been built in the estuaries along the Jiangsu coastline to control saline water intrusion and river flooding (Zhu et al., 2017). Damming at these rivers in the Jiangsu Coast in recent decades has reduced sediment input, even though the impact on the middle coast may be much smaller than the reduction of sediment supply from the abandoned Yellow River delta. The impact of damming of these rivers is more profound on the local mudflat (He et al., 2015) and on the ecosystem near these rivers' mouths than that in the middle coast. In addition, damming in the Yangtze River basin has greatly reduced its overall sediment flux into the ocean (Gao and Wang, 2008). While the Yangtze River provides a very limited portion of sediment to the central Jiangsu mudflat, it could have a large impact on the southern part of the Jiangsu Coast and the impact would be more dramatic during the summer season when the sediment plume has the potential to go northward (Bai et al., 2014; Yuan et al., 2015).

Additional human intervention arises from the digging of coastal sand, as well as the dredging in the navigational channels for major ports. For example, intensive dredging is proposed to deepen the navigational channel for Dafeng Port, aiming to increase its mean water depth from 15 to 16 m. The deepening of the channel affects not only the longitudinal sediment transport but also the lateral sediment transport, both of which would contribute to morphological change in the long term.

5 Weakness of Current Studies on Jiangsu Mudflat

One of the major weaknesses is the cross-scale numerical modeling in the Jiangsu Coast. With the development of numerical modeling systems and the increase of computing capability, numerical models have been widely applied in the coastal ocean study. The tidal dynamics of the Bohai Sea, Yellow Sea, and East China Sea (BYECS) have been extensively investigated by numerous numerical studies using either structured or unstructured grid modeling systems. Here, we highlight some numerical studies with significant importance for our understanding of the dynamics in the Jiangsu Coast. Zhu and Chang (2001) found the convergent tidal current exists even without the constraining of the morphology of radial sand ridges, supporting the hypothesis that the radial sand ridges were formed mainly by the converging tidal current during the period when there were sufficient sediments. Uehara et al. (2002) simulated the tidal dynamics of the East China Sea at 6 ka and 10 ka using paleocoastline bathymetry data and found that there was no amphidrome at 10 ka in the Yellow Sea. Their simulations suggested the tidal dynamics in the basin-wide scale had little variation since 6 ka. Song et al. (2013) and Zhu et al. (2016) studied the impact of extensive reclamation on tidal dynamics and demonstrated that reclamation affects both tidal amplitude and tidal asymmetry, especially in the nearshore area. At the same time, there were increasing efforts on sediment transport simulation. Zhu and Chang (2000) simulated the suspended sediment transport and bed load transport in the BYECS. Xing et al. (2012) used a nested unstructured grid model to simulate the tidal and sediment dynamics in the radial sand ridge area, addressing the seasonality of sediment concentration and transport. Zhang et al. (2013a) calculated sediment flux into the sand ridges area from different directions. Xu et al. (2016) revealed the mechanisms causing the morphological difference between northern and southern sand ridges. Pang et al. (2016) simulated the sediment accumulation in the BYECS and highlighted the significant influence of ocean currents on sediment accumulation, especially on the formation of the mud patch in the Yellow Sea.

However, the development of numerical modeling in this region is hampered by multiple factors, including the following:

a) Lack of high-quality bathymetry. Most of the modeling works were based on the bathymetry measured 30 years ago (Zhu and Chang, 2001; Xing et al., 2012; Du and Wang, 2014; Zhu et al., 2016). Strong sedimentation (erosion and deposition) in this region have changed the bathymetry significantly for the past few decades. Using the old bathymetry while calibrating the model with observational data from recent years will inevitably introduce a considerable amount of error. Quick shifting of tidal channels and intensive sedimentation, especially during typhoon events, make updating of the bathymetry extremely difficult. The bathymetry mapping is also hindered by the large sand ridges. Use of advanced remote sensing techniques, such as satellite and high-resolution LIDAR, might be the solutions to this problem. In addition, satellite data would provide reliable sediment concentration data for model calibration (Wang et al., 2011).

b) Lack of updated sediment types and their grain sizes over a large domain. The latest available map of sediment type is provided by Saito and Yang (1995). Sediment modeling requires an accurate initial composition of the bottom sediment and initial thickness of the active surface sediment layer. Sediment grain size may vary spatially, even within a channel.

c) Lack of long-term water-level measurements at the central Jiangsu Coast between the abandoned Yellow River mouth and Lvsi. Since the significant extension seaward of the coastline, previously constructed tidal gauges along the depositional coast were abandoned and long-term water-level data are usually not available for model calibration and verification.

d) The resuspension and deposition of cohesive sediment, particularly in the tidal flat, have not been well-understood.

Additionally, the great impact from the salt marsh on the accumulation of fine sediment has not been well-parameterized into the sediment modeling framework.

e) A cross-scale modeling work is required, as the hydrodynamics are controlled by remote large-scale tidal waves and interact greatly with the local small-scale morphology. Therefore, appropriate model domains should cover a much larger area than the sand ridge area, which requires a lot of computational resources. Thus, a structured grid model has a limitation to capture well the deep channel and vast tidal flats in the sand ridge area. Even when using the unstructured grid model, a 2D sediment model is generally used, while a 3D sediment model is rarely applied. With the increase of computing resources and advances in parallel computing technique, numerical simulation of the complicated sediment dynamics and morphological processes are feasible the near future.

Despite tremendous studies in this region, there are still numerous questions not clearly answered. For instance, there is a lack of reliable recent sediment budget information in this region. How much sediment comes from the north, from the south, and from offshore? How does the sediment flux change with season? Both the suspended sediment and bedload sediment contribute to the total sediment budget. Even though numerical model simulations have been conducted (e.g., Zhang et al., 2013a), their results are challenged by the accuracy of the bathymetry and the initial sediment composition. Significant discrepancies in the sediment budget can be found among different studies. Considering the decades-long erosion in the outer sand ridges, it is interesting to know how long the seaward advance of the mud-flat will continue.

Event-driven erosion and deposition are not well-documented. Storms have long been recognized as agents of geomorphic change to coastal wetlands (Ren et al., 1983; Deng et al., 1997; Draut et al., 2005; Cahoon, 2014). Frequent typhoons during summer and cold wave events during winter have great impacts on the sedimentation and tidal-flat profile (Fan et al., 2006). There are substantial differences between the responses of sand- and mud-dominated coastal systems to storms (Draut et al., 2005). Storm's impact on morphologic change is profound in many coastal systems. For example, Andersen and Pejrup (2001) found 40% of the annual sediment accumulation in the Danish Wadden Sea mudflat happened during several storms.

The biogeochemical processes (e.g., N and P cyclings) in the mudflat have not been well-documented. Biogeochemical processes are sensitive to external perturbation (e.g., salt-marsh migration), rapid change of climate (e.g., sea-level rise), and human intervention (e.g., damming, reclamation, and aquaculture). In addition, the contaminants discharged by chemical industries are a major concern for the safety of seafood. The contaminants are likely to be redistributed by tidal currents and sediment transports. A coupling of contaminant decomposition processes and sediment transport is necessary to understand the fate of the contaminants.

The future of the Jiangsu mudflat is unknown, as it is influenced by climate, sea-level rise, and human intervention, all of which have a lot of uncertainty. Coastal morphology continually adapts toward an equilibrium status as sea level rises. Marshes struggle to keep pace with the accelerated sea-level rise, which rely on sediment accumulation and the availability of suitable uplands for migration (Passeri et al., 2015). Whether the Jiangsu mudflat will keep pace with the sea-level rise and whether the sediment inputs from the offshore sand ridge and abandoned Yellow River delta are sufficient for maintaining the mudflat in the future are interesting topics to be investigated.

6 Summary

The mudflat of the Jiangsu Coast is influenced by a variety of natural and anthropogenic processes, and has been experiencing a rapid change over the past few decades. Different from the common mudflats found near large river mouths, the Jiangsu mudflat is controlled by unique hydrodynamics, intensive accumulation rate induced by salt marsh, sheltered environment, and sufficient sediment source from the offshore sand ridges and the ancient river deltas.

The interaction between salt marsh and sedimentation processes is particularly interesting, especially under the strong influences of other natural and anthropogenic processes. Salt marsh (esp., *S. alterniflora*) has a profound effect on the fine-sediment accumulation in the Jiangsu mudflat, enhancing the sediment accumulation rate by multiple times and contributing to the continuous accretion and seaward advancing of the mudflat. The salt marsh provides large new land to be reclaimed. Human activities, particularly reclamation, have great potential to alter the natural sedimentation and the sustainability of the wetland, most likely in a negative way.

Extensive geological samplings, hydrodynamics measurements, and numerical simulations were conducted in the past few decades, which have greatly improved our understanding of the past and present conditions of the muddy coast. Without major river sediment input and depending on the remote sediment sources, the muddy coast of Jiangsu is sensitive to the change of ocean circulations, coastal currents, wind field, and global climate. With the development of advanced numerical modeling and remote sensing techniques, a better understanding of the physical, ecological, and morphological dynamics is achievable in the near future.

Acknowledgments

We thank J. H. Gao, Y. Yang, and M. L. Li for their help during the preparation of this manuscript. We thank Prof. Xiao Hua Wang for his careful review and constructive comments on the manuscript. We thank Mac Sisson for his editing and proofreading. Financial support for this study was provided by the National Natural Science Foundation of China (41625021 and 41376044) and the Jiangsu Special Program for Science and Technology Innovation (HY2017-2).

References

Allen, J.R.L., 2000. Morphodynamics of Holocene salt marshes: a review sketch from the Atlantic and southern North Sea coasts of Europe. Quat. Sci. Rev. 19, 1155–1231.

An, S.Q., Gu, B.H., Zhou, C.F., et al., 2007. Spartina invasion in China: implications for invasive species management and future spartina invasion in China. Weed Res. 47, 183–191.

Andersen, T.J., Pejrup, M., 2001. Suspended sediment transport on a temperate, microtidal mudflat, the Danish Wadden Sea. Mar. Geol. 173, 69–85.

Anthony, E.J., Gardel, A., Gratiot, N., Proisy, C., Allison, M.A., 2010. Earth-science reviews the Amazon-in fluenced muddy coast of South America: a review of mud-bank–shoreline interactions. Earth-Sci. Rev. 103 (3–4), 99–121.

Bai, Y., He, X., Pan, D., Chen, C.A., Kang, Y., Chen, X., Cai, W., 2014. Summertime Changjiang River plume variation during 1998-2010. J. Geophys. Res. Oceans 119, 1998–2010.

Bednarek, A.T., 2001. Undamming rivers: a review of the ecological impacts of dam removal. Environ. Manag. 27 (6), 803–814.

Bostock, J., McAndrew, B., Richards, R., et al., 2010. Aquaculture: global status and trends. Philos. Trans. R. Soc. Lond. B Biol. Sci. 365 (1554), 2897–2912.

Cahoon, D., 2014. A review of major storm impacts on coastal wetland elevations. Estuaries 29, 889–898.

Chen, C., 1990. Change trend of erosion and deposition on the mudflat from the Guanhe to the Changjiang River estuary. Mar. Sci. 3, 11–16.

Chen, Y., Gao, S., Jia, J., Wang, A., 2005. Tidal flat ecological changes by transplanting Spartina anglica and Spartina alterniflora, northern Jiangsu coast. Oceanologia et Limnologia Sinica 36 (5), 396–403.

Cooper, N.J., 2005. Wave dissipation across intertidal surfaces in the wash tidal inlet, eastern England. J. Coast. Res. 21 (1), 28–40.

Deng, W., Yang, G., Wu, X., 1997. Studies of storm deposits in China: a review. Cont. Shelf Res. 17 (13), 1645–1658.

Draut, A.E., Kineke, G.C., Huh, O.K., Grymes, J.M., Westphal, K.A., Moeller, C.C., 2005. Coastal mudflat accretion under energetic conditions, Louisiana chenier-plain coast, USA. Mar. Geol. 214, 27–47.

Dronkers, J., 1986. Tidal asymmetry and estuarine morphology. Neth. J. Sea Res. 20, 117–131.

Du, J., Wang, Y.P., 2014. Evolution simulation of radial sand ridges in the southern Yellow Sea. J. Nanjing Univ. 50, 636–645.

Fan, D., Guo, Y., Wang, P., Shi, J.Z., 2006. Cross-shore variations in morphodynamic processes of an open-coast mudflat in the Changjiang Delta, China: with an emphasis on storm impacts. Cont. Shelf Res. 26, 517–538.

Foster, N.M., Hudson, M.D., Bray, S., Nicholls, R.J., 2013. Intertidal mud flat and salt marsh conservation and sustainable use in the UK: a review. J. Environ. Manag. 126, 96–104.

Fu, M., Zhu, D., 1986. The sediment sources of the offshore submarine sand ridge field of the coast of Jiangsu Province. J. Nanjing Univ. Nat. Sci. Ed. 22 (3), 536–544.

Gao, S., Wang, Y.P., 2008. Changes in material fluxes from the Changjiang River and their implications on the adjoining continental shelf ecosystem. Cont. Shelf Res. 28, 1490–1500.

Gao, S., Du, Y.F., Xie, W.J., Gao, W.H., Wang, D.D., Wu, X.D., 2014. Environment-ecosystem dynamic processes of Spartina alterniflora salt-marshes along the eastern China coastlines. Sci. China Earth Sci. 57 (11), 2567–2586.

Green, M.O., Coco, G., 2013. Reviews of wave-driven sediment resuspension and transport in estuaries. Rev. Geophys. 52, 77–117.

Guan, B.X., 1994. Patterns and structures of the currents in Bohai, Huanghai and East China Seas. In: Zhou, D. et al. (Eds.), Oceanology of China Seas. vol. 1. Kluwer Academic Publishers, Dordrecht, pp. 17–26.

He, X.M., 1979. Coastal geomorphology of the Jiangsu Province and radial sand ridges in offshore zone (in Chinese). In: Jiangsu Science Committee (Ed.), Technical Report of Observation on Coastal Zone and Tidal Flat Resources of the Jiangsu Province, pp. 27–40.

He, X., Hu, T., Wang, Y., Zou, X., Shi, X., 2010. Seasonal distributions of hydrometeor parameters in the offshore sea of Jiangsu. Mar. Sci. 34 (9), 44–54.

He, X., Wang, Y.P., Zhu, Q., Zhang, Y., Zhang, D., Zhang, J., Yang, Y., Gao, J., 2015. Simulation of sedimentary dynamics in a small-scale estuary: the role of human activities. Environ. Earth Sci. 74, 869–878.

Lee, H.J., Chao, S.Y., 2003. A climatological description of circulation in and around the East China Sea. Deep Sea Res. II Top. Stud. Oceanogr. 50 (6–7), 1065–1084.

Li, C.X., Zhang, J.Q., Fan, D.D., Deng, B., 2001. Holocene regression and the tidal radial sand ridge system formation in the Jiangsu coastal zone, East China. Mar. Geol. 173 (1–4), 97–120.

Li, B., Liao, C., Zhang, X., et al., 2009. Spartina alterniflora invasions in the Yangtze River estuary, China: an overview of current status and ecosystem effects. Ecol. Eng. 35, 511–520.

Liu, J.P., Xu, K.H., Li, A.C., Milliman, J.D., Velozzi, D.M., Xiao, S.B., Yang, Z.S., 2007. Flux and fate of Yangtze River sediment delivered to the East China Sea. Geomorphology 85, 208–224.

McBride, R.A., Taylor, M.J., Byrnes, M.R., 2007. Coastal morphodynamics and Chenier-plain evolution in southwestern Louisiana, USA: a geomorphic model. Geomorphology 88 (3–4), 367–422.

McCave, I.N., 1970. Deposition of fine-grained suspended sediment from tidal currents. J. Geophys. Res. 75 (21), 4151–4159.

Naimie, C.E., Ann Blain, C., Lynch, D.R., 2001. Seasonal mean circulation in the Yellow Sea—a model-generated climatology. Cont. Shelf Res. 21 (6–7), 667–695.

Pang, C., Li, K., Hu, D., 2016. Net accumulation of suspended sediment and its seasonal variability dominated by shelf circulation in the yellow and East China seas. Mar. Geol. 371, 33–43.

Passeri, D.L., Hagen, S.C., Medeiros, S.C., Bilskie, M.V., Alizad, K., Wang, D., 2015. Earth's future special section: the dynamic effects of sea level rise on low-gradient coastal landscapes: a review. Earth's Future 3, 159–181.

Ren, M., Zhang, R., Yang, J., Zhang, D., 1983. The influence of storm tide on mud plan coast-with special reference to Jiangsu Province. Mar. Geol. Quat. Geol. 3 (4), 1–23.

Ren, M., Zhang, R., Yang, J., 1984. Sedimentation on tidal mud flat in Wanggang area, Jiangsu Province, China. Mar. Sci. Bull. 3 (1), 40–54.

Ren, M.-e. (Ed.), 1986. Comprehensive Investigation of the Coastal Zone and Tidal Land Resources of Jiangsu Province. Ocean Press, Beijing.

Saito, Y., Yang, Z., 1995. Historical change of the Huanghe (Yellow River) and its impact on the sediment budget of the East China Sea. In: Tsunogai, S. et al. (Eds.), Global Fluxes of Carbon and Its Related Substances in the Coastal Sea-Ocean-Atmosphere System, pp. 7–12.

Shen, Y., Feng, N., Zhou, Q., Liu, Y., Chen, Z., 2006. The status and its influence of reclamation on Jiangsu coast. Mar. Sci. 30 (10), 39–43.

Shi, B., Wang, Y.P., Yang, Y., Li, M., Li, P., Ni, W., et al., 2015. Determination of critical shear stresses for erosion and deposition based on in situ measurements of currents and waves over an intertidal mudflat. J. Coast. Res. 31 (6), 1344–1356.

Shi, B., Wang, Y.P., Du, X., Cooper, J.R., Li, P., Li, M.L., Yang, Y., 2016. Field and theoretical investigation of sediment mass fluxes on an accretional coastal mudflat. J. Hydro Environ. Res. 11, 75–90.

Shi, Z., Chen, J.Y., 1996. Morphodynamics and sediment dynamics on intertidal mudflats in China (1961-1994). Cont. Shelf Res. 16 (15), 1909–1926.

Song, D., Wang, X.H., Zhu, X., Bao, X., 2013. Modeling studies of the far-field effects of tidal flat reclamation on tidal dynamics in the East China seas. Estuar. Coast. Shelf Sci. 133, 147–160.

Tamura, T., Horaguchi, K., Saito, Y., Nguyen, V.L., Tateishi, M., Ta, T.K.O., Watanabe, K., 2010. Monsoon-influenced variations in morphology and sediment of a mesotidal beach on the Mekong River delta coast. Geomorphology 116 (1–2), 11–23.

Uehara, K., Saito, Y., Hori, K., 2002. Paleotidal regime in the Changjiang (Yangtze) estuary, the East China Sea, and the Yellow Sea at 6 ka and 10 ka estimated from a numerical model. Mar. Geol. 183 (1–4), 179–192.

Wan, Y., Zhang, Q., 1985. The source and movement of sediments of radiating sand ridges off Jiangsu coast. Oceanologia et Limnologia Sinica 16 (5), 392–399.

Wan, S., Qin, P., Liu, J., Zhou, H., 2008. The positive and negative effects of exotic Spartina alterniflora in China. Ecol. Eng. 35, 444–452.

Wang, Y., Aubrey, D.G., 1987. The characteristics of the China coastline. Cont. Shelf Res. 7 (4), 329–349.

Wang, Y., Zhu, D., 1990. Tidal flats of China. Quat. Sci. 4, 291–300.

Wang, Y., 2000. The mudflat system of China. Can. J. Fish. Aquat. Sci. 40, 160–171.

Wang, A., Gao, S., Jia, J., Pang, S., 2005. Contemporary sedimentation rate on salt marshes at Wanggang, Jiangsu, China. Acta Geograph. Sin. 60 (1), 61–70.

Wang, Q., An, S., Ma, Z., Zhao, B., Chen, J., Li, B., 2006. Invasive Spartina alterniflora: biology, ecology and management. Acta Phytotaxonomica Sinica 44 (5), 559–588.

Wang, X.H., Qiao, F., Lu, J., Gong, F., 2011. The turbidity maxima of the northern Jiangsu shoal-water in the Yellow Sea, China. Estuar. Coast. Shelf Sci. 93 (3), 202–211.

Wang, Y.P., Gao, S., Jia, J., Tompson, C.E.L., Gao, J., Yang, Y., 2012a. Sediment transport over an accretional intertidal flat with influences of reclamation, Jiangsu coast, China. Mar. Geol. 291–294, 147–161.

Wang, Y., Zhang, Y., Zou, X., Zhu, D., Piper, D., 2012b. The sand ridge field of the south Yellow Sea: origin by river-sea interaction. Mar. Geol. 291–294, 132–146.

Xie, W., Gao, S., 2009. The macrobenthos in Spartina alterniflora salt marshes of the Wanggang tidal-flat, Jiangsu coast, China. Ecol. Eng. 35, 1158–1166.

Xing, F., Wang, Y.P., Wang, H.V., 2012. Tidal hydrodynamics and fine-grained sediment transport on the radial sand ridge system in the southern Yellow Sea. Mar. Geol. 291–294, 192–210.

Xiong, J., Wang, X.H., Wang, Y.P., et al., 2017. Mechanisms of maintaining high suspended sediment concentration over tide-dominated offshore shoals in the southern Yellow Sea. Estuar. Coast. Shelf Sci. 191, 221–233.

Xu, K., Milliman, J.D., Li, A., Liu, J.P., Kao, S., Wan, S., 2009. Yangtze- and Taiwan-derived sediments on the inner shelf of East China Sea. Cont. Shelf Res. 29 (18), 2240–2256.

Xu, F., Tao, J., Zhou, Z., Coco, G., Zhang, C., 2016. Mechanisms underlying the regional morphological differences between the northern and southern radial sand ridges along the Jiangsu Coast, China. Mar. Geol. 371, 1–17.

Yanagi, T., Morimoto, A., Ichikawa, K., 1997. Seasonal variation in surface circulation of the East China Sea and the Yellow Sea derived from satellite altimetric data. Cont. Shelf Res. 17 (6), 655–664.

Yang, S., Chen, J., 1994. The role of vegetation in mud coast processes. Oceanologia et Limnologia Sinica 25 (6), 631–635.

Yang, G., Shi, Y., Ji, Z., 2002. The morphological response of typical mud flat to sea level change in Jiangsu Coastal plain. Acta Geograph. Sin. 57 (1), 76–84.

Yang, B., Feng, W.B., Zhang, Y., 2014. Wave characteristics at the south part of the radial sand ridges of the southern Yellow Sea. China Ocean Eng. 28 (3), 317–330.

Yu, Z., Zhang, Y., 1994. The erosion process of open mud beach and its protection along the coast of northern Jiangsu. Acta Geograph. Sin. 49, 149–157.

Yuan, R., Wu, H., Zhu, J., Li, L., 2015. The response time of the Changjiang plume to river discharge in summer. J. Mar. Syst. 154, 82–92.

Zhang, R., 1986. Characteristics of tidal current and sedimentation of suspended load on tidal mud flat in Jiangsu Province. Oceanologia et Limnologia 17 (3), 235–245.

Zhang, R., 1992. Suspended sediment transport processes on tidal mud flat in Jiangsu Province, China. Estuar. Coast. Shelf Sci. 90, 225–233.

Zhang, R., Lu, L., Wang, Y., 2002. The mechanism and trend of coastal erosion of Jiangsu Province in China. Geogr. Res. 21 (4), 469–478.

Zhang, R.S., Shen, Y.M., Lu, L.Y., Yan, S.G., Wang, Y.II., Li, J.L., Zhang, Z.L., 2004. Formation of Spartina alterniflora salt marshes on the coast of Jiangsu Province, China. Ecol. Eng. 23, 95–105.

Zhang, C., Chen, J., Lin, K., Ding, X., Yuan, R., Kang, Y., 2011. Spatial layout of reclamation of coastal tidal flats in Jiangsu Province. J. Hohai Univ. 3 (2), 206–212.

Zhang, C., Yang, Y., Tao, J., et al., 2013a. Suspended sediment fluxes in the radial sand ridge field of south Yellow Sea. J. Coast. Res. 65, 624–629.

Zhong, C.X., Zhuo, R.Z., 1985. Twenty two years of spartina algilica Hubbard in China. J. Nanjing Univ., 31–35.

Zhou, H., Liu, J., Qin, P., 2008. Impacts of an alien species (Spartina alterniflora) on the macrobenthos community of Jiangsu coastal inter-tidal ecosystem. Ecol. Eng. 5, 521–528.

Zhu, D., Ke, X., Gao, S., 1986. Tidal flat sedimentation of Jiangsu coast. J. Oceanogr. Huanghai Bohai Seas 4 (3), 19–27.

Zhu, Y., Chang, R., 2000. Preliminary study of the dynamic origin of the distribution pattern of bottom sediments on the continental shelves of the Bohai Sea, Yellow Sea and East China Sea. Estuar. Coast. Shelf Sci. 51 (5), 663–680.

Zhu, Y., Chang, R., 2001. On the relationships between the radial tidal current field and the radial sand ridges in the southern Yellow Sea: a numerical simulation. Geo-Mar. Lett. 21 (2), 59–65.

Zhu, Q., Wang, Y.P., Ni, W., et al., 2016. Effects of intertidal reclamation on tides and potential environmental risks: a numerical study for the southern Yellow Sea. Environ. Earth Sci. 75 (23), 1–17.

Zhu, Q., Wang, Y.P., Gao, S., Zhang, J., Li, M., Yang, Y., Gao, J., 2017. Modeling morphological change in anthropogenically controlled estuaries. Anthropocene 17, 70–83.

Zhuo, R.Z., Xu, G.W., 1985. A note on trial planting experiments of Spartina alterniflora. J. Nanjing Univ. 40 (2), 353–354.

Zuo, P., Liu, C.H., 2008. Analysis on the impact of the exotic plant species in China's coastal zone: *Spartina anglica* and *Spartina alterniflora* as example. Ocean Develop. Manage. 12, 107–122.

Further Reading

Gao, S., Wang, D., Yang, Y., et al., 2015. Holocene sedimentary systems on a broad continental shelf with abundant river input: process-product relationships. In: Clift, P.D. et al. (Eds.), River-Dominated Shelf Sediments of East Asian Seas.

Zhang, X., Yan, C., Pan, X.U., Dai, Y., Yan, W., Ding, X., et al., 2013b. Historical evolution of tidal flat reclamation in the Jiangsu coastal areas. Acta Geograph. Sin. 68 (11), 1549–1558.

4

CHANGJIANG ESTUARY

Jianrong Zhu, Hui Wu, Lu Li, Cheng Qiu
State Key Laboratory of Estuarine and Coastal Research, East China Normal University, Shanghai, China

CHAPTER OUTLINE

1 **Introduction** 51
2 **Numerical Model** 53
3 **Dynamic Factors Controlling the Water Movement in the Changjiang Estuary** 55
   3.1 Runoff 55
   3.2 Tide 56
   3.3 Wind 57
   3.4 Residual Current 60
4 **Saltwater Intrusion** 62
   4.1 Impact of Wind 64
   4.2 Impact of Sea Level Rise 64
   4.3 Impact of Major Projects 65
5 **Hypoxia off the River Mouth** 67
6 **Summary** 71
References 73
Further Reading 75

1   Introduction

The Changjiang, also known as the Yangtze River, has the fourth largest mean sediment load and fifth largest water discharge of rivers worldwide, delivering $\sim 9.24 \times 10^{11}\,\mathrm{m^3/yr}$ of freshwater to the Yellow Sea and East China Sea (Shen et al., 2003). The Changjiang Estuary has a 90-km-wide river mouth, and is characterized by multiple bifurcations due to sand islands (Fig. 1). Downstream from Xuliujing, the estuary is divided by Chongming Island into the South Branch (SB) and North Branch (NB). The SB and its lower reaches form the main channel of the Changjiang

Sediment Dynamics of Chinese Muddy Coasts and Estuaries. https://doi.org/10.1016/B978-0-12-811977-8.00004-2

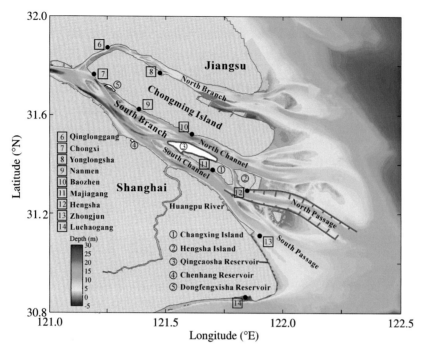

Fig. 1 Map of the Changjiang Estuary. Key geographic locations and tide gauges are *numbered* and marked by *circles* and *triangles*.

and contain most of the river discharge, while the NB is heavily silted. The lower SB is bifurcated into the South Channel (SC) and the North Channel (NC) by Changxing Island and Hengsha Island. Finally, the SC is bifurcated into the South Passage (SP) and the North Passage (NP) by Jiuduan tidal flat. Thus, four outlets, separated by extensive intertidal flats, discharge from the Changjiang into the East China Sea.

The Changjiang Estuary is unique among well-studied estuaries, due to its large size, high river discharge, considerable tidal range, and relatively high suspended sediment concentration. It also has an extremely dynamic hydrological environment due to runoff, tide, wind, mixing, topography, and continental shelf current outside the river month, which are the main dynamic controlling factors on hydrodynamic processes in the Changjiang Estuary (Wu et al., 2006; Xiang et al., 2009; Li et al., 2012). Both river discharge and wind have strong seasonality. Wind is controlled primarily by the monsoons, which bring weak southerly winds in summer and strong northerly winds in winter. Tides exhibit a medium tidal range and are the most energetic source of water movement in the Changjiang Estuary. The tidal range introduces a significant tidal current, which oscillates and mixes dissolved materials. These dynamics have strong influences on circulation, saltwater intrusion, plume dispersal, sedimentary erosion/deposition, and other biogeochemical processes (Zhang, 2015).

In this chapter, we describe and analyze the hydrodynamics, saltwater intrusion and hypoxia in the Changjiang Estuary. In Section 2, we present the numerical model that is applied to simulate the dynamic processes and saltwater intrusion in the estuary. In Section 3, we analyze and simulate the runoff, tide, wind-driven current, and total residual current. In Section 4, we simulate and analyze saltwater intrusion. In Section 5, we briefly describe the hypoxia off the Changjiang mouth; and in Section 6, a summary is provided.

2   Numerical Model

Under the assumption of incompressibility, as well as Boussinesq and hydrostatic approximations, and using the horizontal nonorthogonal curvilinear and vertically stretched sigma coordinate system, ECOM-si was developed based on the Princeton Ocean Model (POM; Blumberg and Mellor 1987; Blumberg 1994) with several improvements (Chen et al., 2001) to address issues related to numerical simulations of water bodies bounded by complicated coastlines. This model incorporates the Mellor-Yamada level-2.5 turbulent closure scheme to provide a time-and-space dependent parameterization of vertical turbulent mixing (Mellor and Yamada 1974, 1982; Galperin et al., 1988). Wu and Zhu (2010) developed the 3rd HSIMT-TVD scheme to solve the advection term in the mass transport equation. This scheme is flux-based with 3rd-order accuracy in space, 2nd-order accuracy in time and no numerical oscillation. A wet/dry scheme was included to characterize the intertidal zone due to tidal excursion, and the critical depth was set to 0.2 m.

The model domain covered the whole Changjiang Estuary, Hangzhou Bay, and adjacent seas from 117.5°E to 125°E and from 27.5°N to 33.7°N (Fig. 2A). The model grid mesh was composed of 307×224 cells horizontally and 10 uniform σ levels vertically. The mesh was designed to fit the coastline, with high resolution grids near the Changjiang mouth, especially near the bifurcation of the SB and NB (Fig. 2B), and near the NP where a deep waterway was maintained for navigation (Fig. 2C). A lower resolution grid was used in open water. The grid resolution ranged from 300 to 600 m in proximity to the river mouth and was 15 km near open water.

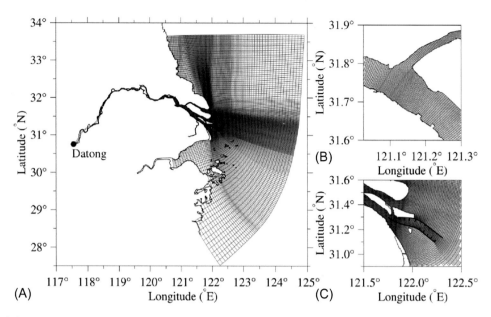

Fig. 2 (A) The numerical model mesh; (B) an enlarged view of the model mesh around the bifurcation of the NB and the SB; and (C) an enlarged view of the model mesh in the NP, showing the location of the deep waterway project.

Derived from the NaoTide dataset (http://www.miz.nao.ac.jp/), the open sea boundary included 16 astronomical constituents: M_2, S_2, N_2, K_2, K_1, O_1, P_1, $Q1$, MU_2, NU_2, T_2, L_2, $2N_2$, J_1, M_1, and OO_1. Daily or monthly mean river discharge recorded at Station Datong was used in the model as the river boundary condition. Wind field, used to calculate sea surface momentum, was simulated by the Weather Research Forecast Model (WRF), or from the weather station at the Chongming eastern shoal. We used the National Center for Environmental Prediction (NCEP) reanalysis dataset with a spatial resolution of $0.5° \times 0.5°$ and a temporal resolution of 6h as initial and boundary conditions of the WRF model.

The velocities and elevation were initially set to zero. The initial salinity distribution was derived from the Ocean Atlas in the Huanghai Sea and East China Sea (Hydrology) (Editorial Board for Marine Atlas 1992) outside the Changjiang mouth, and from the observed data in the estuary in recent years. Because salinity dominates the density variability in the Changjiang Estuary, water temperature was set to a constant in the model. A detailed description of the model configuration and validation can be found in Wu et al. (2010), Li et al. (2012), and in Qiu and Zhu (2013).

3 Dynamic Factors Controlling the Water Movement in the Changjiang Estuary

The water movement in the Changjiang Estuary is controlled mainly by runoff, tide, and wind. We analyzed each dynamic factor and simulated the current driven by each of individual factors as well as combined dynamic factors using the numerical model described in Section 2.

3.1 Runoff

We collected the measured river discharge data at Station Datong since 1950, which accounts for 94.72% of the total river basin, is the upper tidal limit in dry seasons and is generally used as an upper boundary of the estuary. The river discharge has seasonal variation, which increases from January to July, and decreases from July to December (Table 1). The monthly mean river discharge has the minimum of $11,184 \, m^3/s$ in January, and reaches the maximum of $49,848 \, m^3/s$ in July. The annual mean river discharge is $28,460 \, m^3/s$.

To understand the impact of runoff in the estuary, the model was run only with the river discharge. The river boundary in the model is the location of Datong Hydrographic Station, which is 630 km upstream from the river mouth. We carried out two experiments by monthly mean river discharge of $11,184 \, m^3/s$ in January and of $49,848 \, m^3/s$ in July, respectively. When driven only by river discharge, the runoff flows seaward, with a larger speed of about 0.12 m/s in the SB, 0.08–1.00 m/s in the NC and SC, and 0.05 m/s in the NP and SP in winter in the main channel, and with smaller speeds in the shallow shoals (Fig. 3A). A small part of river discharge flows into the NB, and the runoff is small there, especially in the

Table 1 Monthly mean river discharge (m³/s) of the Changjiang from 1950 to 2016 at Station Datong

Month	1	2	3	4	5	6
Value	11,184	12,058	16,300	24,061	33,488	40,454
Month	7	8	9	10	11	12
Value	49,848	44,149	39,909	32,332	22,552	14,063

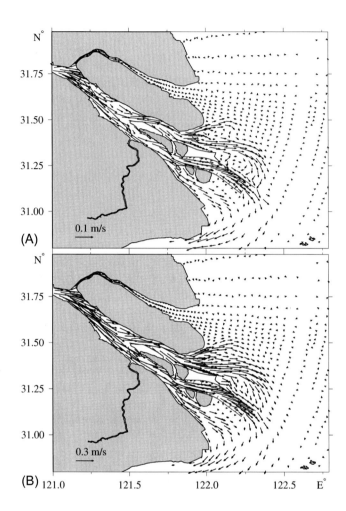

Fig. 3 Simulated depth-averaged runoff driven by river discharge of 11,200 m³/s in January (A) and of 49,900 m³/s in July (B).

lower reaches. The pattern of the runoffs in summer is the same as that in winter, but the velocity is about three times larger than that in winter due to larger river discharge in summer (Fig. 3B).

3.2 Tide

Tides are the most energetic source of water movement in the Changjiang Estuary. To understand the tidal characteristics in the study area, we used tidal data from six hydrological stations: Hengsha, Majiagang, Baozhen, Yonglongsha, Chongxi, and Nanmen (Fig. 1). The data were collected hourly throughout 2009, providing sufficient resolution to reveal the temporal variability of tides in the estuary. We used harmonic analysis to characterize the key tidal constituents.

The tidal elevation not only exhibited semidiurnal variability but also had a significant fortnightly variability (Fig. 4). Water levels in December were generally lower than those in September due to the reduction in river discharge. Monthly mean water levels at the six stations all showed significant seasonal variability, reflecting the variability in river discharge, with the highest value around August and the lowest value in January. Monthly mean water levels at Station Yonglongsha were lower than those at the other five stations, because the tidal flats in the upper reaches of the NB restrained the runoff into the NB.

Fig. 4 also distinctly illustrates the temporal variation of the tidal range at the six stations in September and December 2009. For example, at Station Hengsha, the maximum tidal range of 360 cm occurred during a spring tide on September 20, which was only 130 cm on September 27 during a neap tide. The difference between the spring and neap tidal ranges was primarily due to differences in high-tide water level, which was more variable during a neap tide. The maximum tidal range in December was smaller than that in September. The fortnightly variability of the tidal range was weaker in December than in September. The seasonal and fortnightly variations in September and December at the other five stations were comparable to those at Station Hengsha (Fig. 4B–D). Note that the tidal range at Station Yonglongsha exceeded 500 cm in September and reached 400 cm in December. This range was considerably larger than those at the other stations due to the location of Station Yonglongsha in the NB, where the funnel-shaped topography concentrated flow and the tidal forcing was stronger. Detailed seasonal and spatial variability of tidal range and tidal constituents can be found in Zhu et al. (2015a,b).

To simulate tidal currents, the numerical model was driven only by the tides. At the maximum flood current in spring tide (reference site was Station Baozhen), the flood currents in the NB, SB, SC, middle and upper reaches of NC, SP, and NP flow landward with the speed of 1.0–1.8 m/s, while the tidal currents turn to ebb currents in the lower reaches of the NC, NP and SP due to a larger area of the estuary (Fig. 5A). For the maximum ebb current in spring tide, the ebb currents in the NB, SB, SC, NC, SP, and NP flow seaward with the speed of 1.0–2.0 m/s, while the tidal currents turn to flood currents near the river mouth (Fig. 5B).

3.3 Wind

The prevailing monsoon climate results in strong northerly wind during winter and weak southerly wind during summer. The wind field used in the model was the NCEP reanalysis dataset.

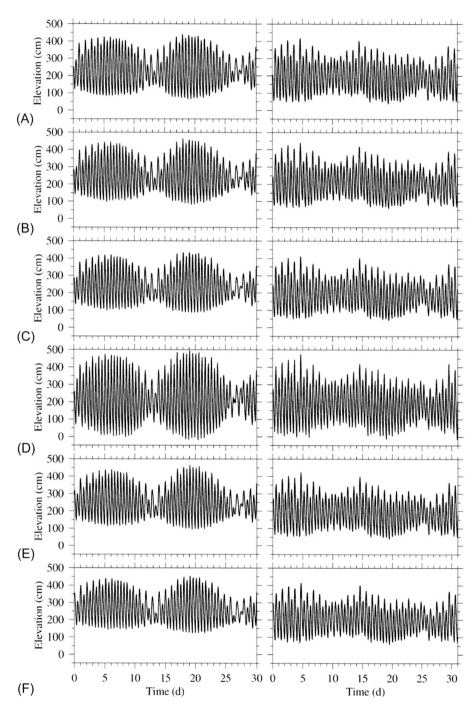

Fig. 4 Temporal variation of the measured water level (tidal elevation) in 2009 at: (A) Station Hengsha, (B) Station Majiagang, (C) Station Baozhen, (D) Station Yonglongsha, (E) Station Nanmen, and (F) Station Chongxi during September *(left panels)* and December *(right panels)*.

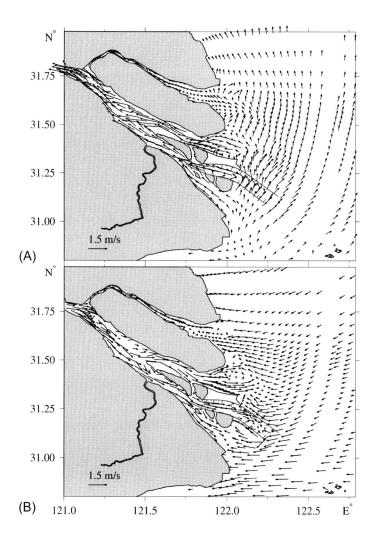

Fig. 5 Tidal current distribution at the maximum flood current (A) and at the maximum ebb current (B) in spring tide.

Fig. 6 shows the monthly mean wind vectors in the Changjiang Estuary in January and July, respectively, and illustrates that northerly wind prevails in winter with a speed of ~6 m/s, and southerly wind prevails in summer with a speed of ~4 m/s near the river mouth.

To simulate the wind-driven current in the estuary, the model was integrated only with the monthly mean wind in January and July, respectively. Forced by the strong monsoon in winter, the wind-driven surface current flowed southward off the river mouth, and flowed landward into the NC, then turned to the SC, and flowed seaward into the NP and SP with a speed of 0.2–0.4 m/s forced by the Ekman transport (Fig. 7A). In summer,

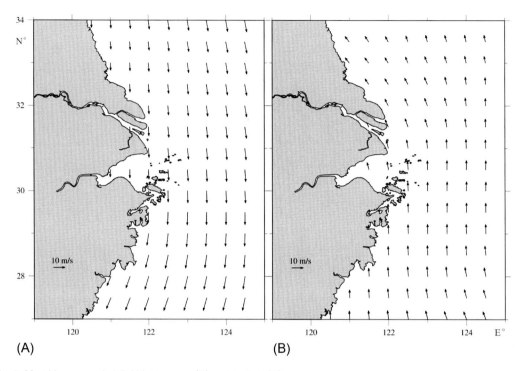

Fig. 6 Monthly mean wind field in January (A) and in July (B).

the flow direction was opposite (Fig. 7B), and the wind-driven current was weaker due to weak summer monsoon.

We used the wave model SWAN to calculate the wave off the Changjiang mouth. Driven by the northerly wind in January (Fig. 6A), the wave propagated from north to south, and the signification wave height was 0.6–1.0 m near the river mouth (Fig. 8A). In contrast, when driven by the southerly wind in July (Fig. 6B), the wave propagated from south to north, and the signification wave height was 0.4–0.5 m near the river mouth (Fig. 8B).

3.4 Residual Current

Residual current, also known as subtidal circulation, governs the long-term transport of conserved materials and sediments. Residual current in an estuary is mainly induced by low-frequency forces, such as runoff, wind, baroclinic gradient, and the tide itself (Pritchard 1956).

Considering the combined effect of river discharge, tide, wind, and baroclinic force induced by the density gradient, the total residual current in the Changjiang Estuary was simulated with

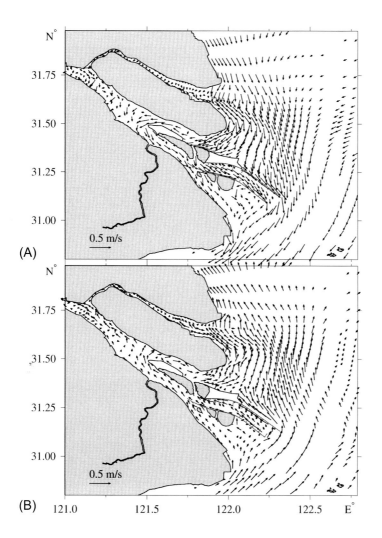

(A)

(B) 121.0   121.5   122.0   122.5   E°

Fig. 7 Wind-driven current driven by the monthly mean wind in January (A) and in July (B).

all these forcing factors included. Fig. 9A shows the winter residual current during spring tide. In the SB, SC, NC, and NP, river discharge causes the residual current to move seaward, which has a higher amplitude in these channels than over the tidal flats. In the sand-bar areas between the SP and NP, water is transported northward across the tidal flats, which is consistent with observational results (Wu et al., 2010). The residual current over the tidal flat east of Chongming Island is northward, due to the tide-induced Stokes drift. In the NB, the residual current is weaker, and there is a net water transport toward the SB. Fig. 9B shows the summer residual current during spring tide. In the SB, SC, NC, and NP, the residual current is seaward and has a larger

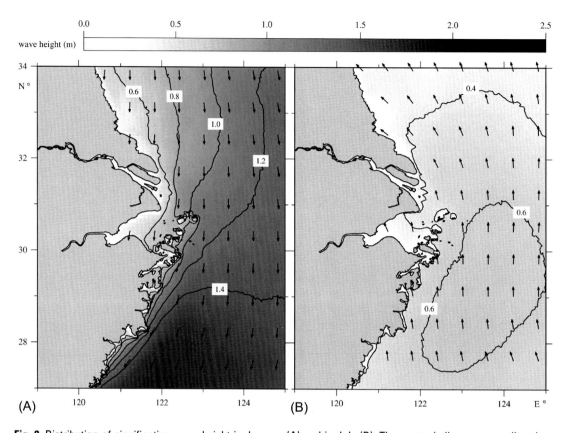

Fig. 8 Distribution of signification wave height in January (A) and in July (B). The vector indicates wave direction.

amplitude in summer than in winter, due to high river discharge in summer. The residual current over the tidal flat east of Chongming Island is northward due to tidal pumping and southerly wind. In the NB, the residual current is weaker and almost seaward, and the net water transport is from the SB into the NB.

4 Saltwater Intrusion

Saltwater intrusion in the Changjiang Estuary is controlled mainly by the river discharge and tide (Shen et al., 2003; Wu et al., 2006; Li et al., 2010; Zhu et al., 2010; Qiu et al., 2012), but is also influenced by wind (Li et al., 2012), topography (Li et al., 2014), river basin and estuarine projects (Zhu et al., 2006; Qiu and Zhu 2013), and sea level rise (Qiu and Zhu 2015). Strong tidal forcing in the NB induces significant subtidal circulation, resulting in a net landward flow when river discharge is low during

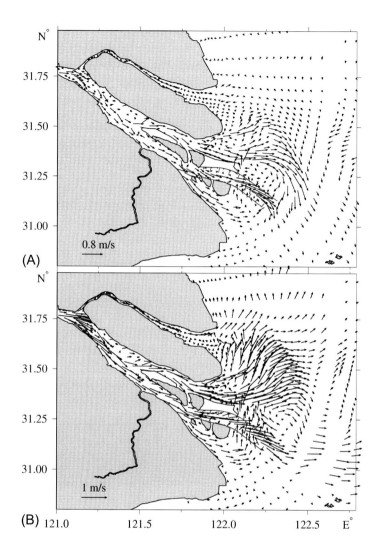

Fig. 9 Distribution of the total residual current during spring tide in January (A) and in July (B).

spring tide (Wu et al., 2006). This residual transport forms a type of saltwater intrusion known as the saltwater-spilling-over (SSO), which is the most striking characteristic of saltwater intrusion in the estuary. During spring tide, the water level rises considerably in the upper reaches of the NB due to its funnel shape, leading to a massive amount of saline water spilling over the shoal into the SB. The saline water spilled into the SB is then transported downstream forced by river discharge, and arrives in the middle reaches of the SB during the subsequent neap tide. This process impacts the Chenhang and Qingcaosha reservoirs and threatens Shanghai's water supply. Considering wind, sea level rise, and the major projects in the river basin and estuary, we simulated and analyzed their impacts on the saltwater intrusion in the Changjiang Estuary with the numerical model described in Section 2.

4.1 Impact of Wind

The observation at the Chongxi gauge station indicated the salinity of saline water spilling over from the NB to the SB increased abnormally during strong northerly wind (Li et al., 2012). The numerical model reproduced the phenomenon of abnormal salinity rise, and showed that the salinity there was significantly reduced if the wind speed was reduced by half. Driven by the monthly mean river discharge of $11,000\,m^3/s$ and northerly wind of $5\,m/s$ from January to February, the model simulated the temporal and spatial variation of saltwater intrusion. The net water (net salt) flux in the upper reaches of the NB is $-531\,m^3/s$ ($-16.00\,t/s$) during spring tide, and $-12\,m^3/s$ ($-2.08\,t/s$) during neap tide. The water and salt spill over from the NB to the SB, which is greater during spring tide than during neap tide. Strong northerly wind enhances the net water and salt fluxes from the NB to the SB, and enhances the wind-driven circulation that flows landward in the NC and seaward in the SC (Fig. 10), which increases the saltwater intrusion in the NC (Li et al., 2012). Wind is an important dynamic factor for saltwater intrusion in the Changjiang Estuary.

4.2 Impact of Sea Level Rise

Global sea level rise has caused great concern among governments and societies with its impact on saltwater intrusion and material transport in estuaries, threatening freshwater habitats and drinking water supplies, and so on (submerged island nations, etc.). Qiu and Zhu (2015) simulated the variations of saltwater intrusion and residual salt transport according to several sea level rise scenarios in a typical year and a dry year. Results from their numerical experiments showed that both the intensity of saltwater intrusion and stratification increase as sea level rises, while the increment is quite distinct in each channel. Furthermore, they showed obvious inter-annual changes with different river discharges. The residual transport of salt was used to analyze the changes of transport process and estuarine circulation pattern. The model results showed that the Stokes drift transport is the major mechanism for the up-estuary salt transport in each channel, while the seaward Eulerian transport is of the same order in magnitude. The flux of SSO from the NB into the SB increases

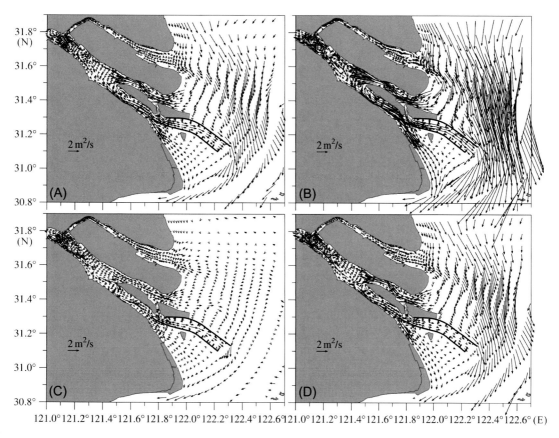

Fig. 10 Pure wind-driven net unit width water flux under the northerly wind of 5 m/s (A), northerly wind of 10 m/s (B), northeasterly wind of 5 m/s (C) and northwesterly wind of 5 m/s (D).

due to stronger Stokes drift transport as sea level rises. The landward salt transport is strengthened in the NC and may further affect the upper reaches under strengthened tidal pumping and vertical shear transport for higher sea level. The overtopping flow affected by tidal pumping is the dominant mechanism for salt exchange between the NP and the SP. This may increase the salt supply from the ocean into the SC.

4.3 Impact of Major Projects

Using the numerical model in Section 2, Zhu et al. (2006) analyzed the impact of the deep waterway project on the saltwater intrusion in the Changjiang Estuary. In the NC, the saltwater intrusion has been alleviated distinctly after the deep waterway project, because the dykes of the project block off the southward drift of the brackish water plumes under the northerly monsoon and the Coriolis force. The saltwater intrusion in the project area is intensified at the upper section and alleviated at the lower section. In the SP, the saltwater intrusion is intensified as the background salinity increased and the river discharge decreased. The deep waterway project has an obvious impact on the saltwater intrusion in the Changjiang Estuary.

The TGR is the largest water conservancy project in the world. It significantly regulates discharge of the Changjiang on a seasonal scale. It stores water in autumn and releases it during the following dry season in winter. Qiu and Zhu (2013) used the numerical model in Section 2 to simulate the seasonal-scale saltwater intrusion around the Changjiang Estuary under the scenarios with and without the TGR regulation. When the river discharge is regulated at a rate of about $4272\,m^3/s$ in autumn when the TGR stores water, the net water flux decreases in each channel. During spring tide, the difference between the residual water transports before and after the TGR began operating is almost landward in the estuary due to the reduced river discharge (Fig. 11A). Salinity increases in the middle NB, indicating that the distance of upstream movement of higher salinity water increased after the TGR began operating. Around the river mouth, the salinity generally increases in the SP, NP and the lower NC to a value of about 1.5. During neap tide, the salinity increases around the river mouth (Fig. 11B). The increment of salinity is larger than 2 in the middle NB and lower NC, and it reaches ~2.5 around the Jiuduan Sandbank.

The seaward residual water transport is augmented during the dry season after the TGR began operating, which means that more fresh water is discharged into the sea, resulting in weaker saltwater intrusion in each channel. During spring tide, the salinity generally decreases in the estuary (Fig. 11C). The net water flux increases to $35\,m^3/s$ in the NB and dilutes the high salinity water in its upper reaches. This leads to a decline in the salt flux that spills over into the SB. Around the river mouth, the salinity in the NP and the SP generally decreases to a value of about 1. In the NC, the salinity also decreases. During neap tide, the salinity difference reaches -2.5 in the upper NB (Fig. 11D), mainly because of the seaward movement of the salinity front. The value of salinity difference is about 1 in the SP and the NP, and it decreases slightly in the NC. So, these results showed that during the autumn season, the TGR advances the timing of saltwater intrusion and slightly increases its intensity. In contrast, as the TGR supplements river discharge during the dry season, saltwater intrusion is suppressed. These model results show that the operation of the TGR is basically favorable for reducing the burden of freshwater supplement in the highly-populated estuarine region.

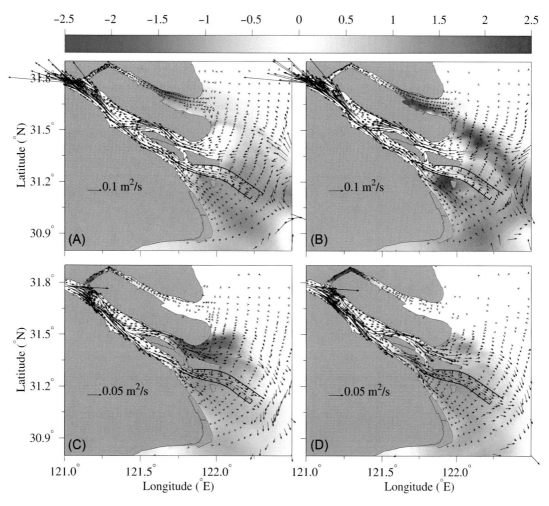

Fig. 11 Differences between vertically averaged residual water transport and salinity before and after the TGR began operating. Comparison is for spring tide (A) and neap tide (B) in September and for spring tide (C) and neap tide (D) in March. The *positive values* mean salinity increase after the TGR, and the *negative values* indicates salinity decrease after the TGR.

5 Hypoxia off the River Mouth

Hypoxia is generally defined as an aquatic dissolved oxygen (DO) concentration of less than 2.0 mg/L (Diaz and Rosenberg 1995), and has been widely observed in various estuarine and coastal regions during the recent decades. Coastal hypoxia can change the natural redox condition of the marine environment, which alters material cycles (Turner et al., 2008; Bianchi and Allison 2009), reduces biodiversity and alters community structures and ecology (e.g., Rabalais et al., 2001). There are two major causes of hypoxia. The first involves oxygen depletion due to the decomposition of large quantities of organic matter from algae in the upper mixed layer (Turner and Rabalais 1994). Excessive nutrient loading often leads to eutrophication (e.g., Hecky and Kilham 1988; Kemp et al., 2005), which results in algal blooms. The second cause involves stratification, which separates the oxygen-rich surface water from the water at depth and limits oxygen exchange (Rosenberg et al., 1991). Estuarine areas are often hypoxic because of the presence of excess nutrients, which can cause algal blooms, and of freshwater input, which can intensify stratification.

Hypoxia has occurred off the Changjiang mouth (e.g., Li et al., 2002; Dai et al., 2006; Wei et al., 2007). Low DO concentration in the region has become more severe in recent decades, with the area of hypoxic zone increasing from $1900 \, km^2$ in 1959 to $15,400 \, km^2$ in 2006 (Zhu et al., 2011). The Changjiang annually empties $9.32 \times 10^{11} \, m^3$ of fresh water into the East China Sea. This input includes significant sediment ($468 \times 10^6 \, t/yr$), nitrate ($6.3 \times 10^6 \, t/yr$), phosphate ($0.13 \times 10^6 \, t/yr$), and dissolved silica ($20.4 \times 10^6 \, t/yr$) loads (Chen et al., 1989). With the rapid development of agriculture and industry in the Changjiang Basin, the amounts of terrestrial nutrients and pollutants transported into the East China Sea have increased over time. This increase has resulted in intensified eutrophication in the adjacent seas and led to severe and frequent algal bloom events. The nitrate concentration in the Changjiang Estuary increased from $20.5 \, \mu mol/L$ in the 1960s, to $59.1 \, \mu mol/L$ in the 1980s and to $80.6 \, \mu mol/L$ between 1990 and 2004 (Zhou et al., 2008). The phosphate concentration increased from $0.59 \, \mu mol/L$ in the 1980s to $0.77 \, \mu mol/L$ between 1990 and 2004. The presence of excessive nutrients can cause eutrophication and stimulate algal blooms, which have been observed with increasing frequency in the area bordering the Changjiang Estuary (Chen et al., 2003; Zhu, 2005; Zhang et al., 2015). Li et al. (2002) reported two regions with depleted DO in the estuary, one in the north and the other in the south. Wang (2009) reported that all the observed hypoxia events off the Changjiang Estuary covered a total area larger than $5000 \, km^2$ after the late-1990s, which suggests an increasing trend in hypoxia.

Based on a large number of observations in the waters off the Changjiang Estuary, Zhu et al. (2015b) studied the distribution of hypoxia and pycnocline in summer. There were frequent hypoxia events, such as the one observed from August 27 to September 3, 2009 (Fig. 12). The DO distribution showed that there was a large area with DO less than $3.0 \, mg/L$ in the bottom layer, and hypoxia existed over an area of $3735 \, km^2$. The distribution of hypoxia generally matched the distribution of pycnocline well.

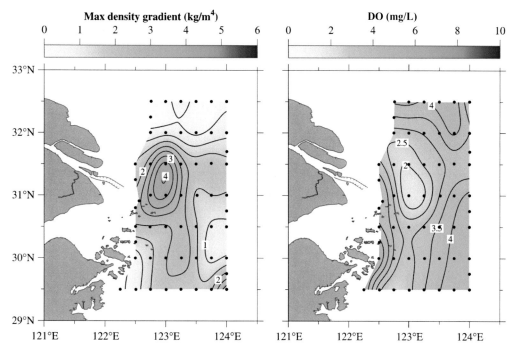

Fig. 12 Horizontal distribution of the maximum density gradient *(left)* and DO distribution in the bottom waters *(right)* during the observation period of August 27 to September 3, 2009.

In addition to the obstruction of vertical oxygen exchange by the pycnocline, the horizontal water exchange in the bottom layer can influence hypoxia. The short residence time of near-bottom waters can result in the dramatic decrease of DO, and the low current speed is favorable for the occurrence of hypoxia. By considering tide, monthly mean river discharge, wind, and open boundary conditions, we numerically simulated the current, salinity and temperature in the East China Sea and Yellow Sea. A detailed description of the configuration of the model used and its validation can be found in Wu et al. (2011). The climatological residual currents in the bottom layer in June, July, August, and October after the tidal current is filtered are presented in Fig. 13. The northward-flowing Taiwan Warm Current occupies the area off the Changjiang mouth in the bottom layer with relatively larger speed along the west slope of the submarine valley. The area of the current speed greater than 6 cm/s is larger in June, becomes smaller in July, is much smaller in August, and becomes larger again in October. Overall, the residual current speed is small (ranging from 2 to 8 cm/s) and is favorable for the occurrence of hypoxia, especially in August, which potentially explains why

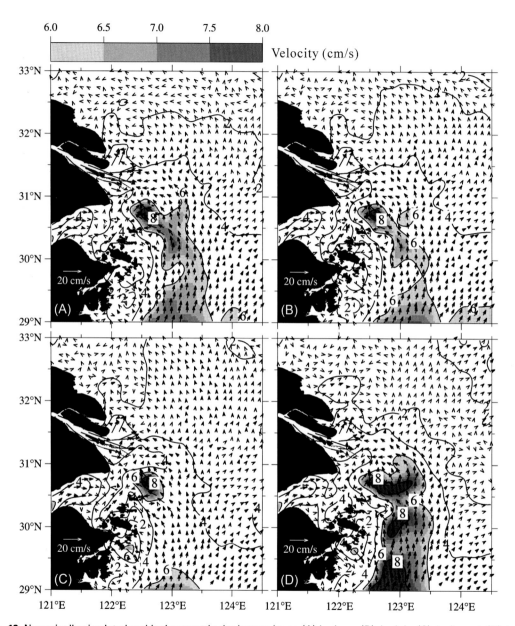

Fig. 13 Numerically simulated residual current in the bottom layer. (A) in June; (B): in July; (C): in August; (D): in October. The contours of current speeds are labeled. The *light* and *dark shaded* areas represent the current speed greater than 6 and 8 cm/s, respectively.

hypoxia often occurs in August. The low climatological residual current in the near-bottom waters prevents lateral advection. Together with strong stratification, the near-bottom waters are nearly isolated from the ambient environment. When excess labile organic matter is imported into these waters (e.g., via sinking of fresh particulate organic matter derived from production above the pycnocline), hypoxia occurs.

6 Summary

The Changjiang is one of the largest rivers in the world. The Changjiang Estuary has an extremely dynamic hydrological environment due to runoff, tide, wind, mixing, topography, and continental shelf current outside the river mouth, which are the main controlling factors on the hydrodynamic processes in the Changjiang Estuary.

The temporal and spatial variation of tides in the estuary was analyzed based on in situ hourly data from tidal gauge stations. Significant semi-diurnal, fortnightly, and seasonal tidal variations were found. The maximum monthly tidal range exhibited two peaks in March and September, and two valleys in June and December each year. As the tidal wave propagated up-estuary, the tidal range became smaller due to bottom friction and runoff.

When driven only by the river discharge of $11,184\,m^3/s$ in January in the model, the simulated runoff flows seaward, with a larger speed of about $0.12\,m/s$ in the SB, 0.08–$1.0\,m/s$ in the NC and SC, and $0.05\,m/s$ in the NP and SP in the main channel in winter, and with smaller speeds in the shoals. A small part of river discharge flows into the NB, and the runoff there is small due to its shallow topography. The pattern of the runoff in summer is the same as that in winter, but the velocity is about three times the one in winter due to the larger river discharge of $49,848\,m^3/s$ in July imposed in the model. The maximum flood and ebb currents are about $2.0\,m/s$. Forced by the strong monsoon in winter, the wind-driven surface current flows southward off the river mouth, flows landward into the NC, then turns to the SC, and flows seaward into the NP and SP with a speed of 0.2–$0.4\,m/s$ due to the Ekman transport. In summer, the situation is opposite due to weaker summer monsoon. When driven by the northerly wind in January, the wave propagates from north to south, and the signification wave height is 0.6–$1.0\,m$ near the river mouth. In contrast, when driven by the southerly wind in July, the wave propagates from south to north, and the signification wave height is 0.4–$0.5\,m$ near the river mouth. Considering the combined effect of river discharge, tide, wind, and baroclinic force induced by the density gradient, the simulated winter residual current during spring tide in the SB, SC, NC, and NP flows seaward mainly due to river discharge, and has a larger magnitude in these channels than over the tidal flats. In the sand-bar areas between the SP and NP, water is transported northward across the tidal flats. The residual current over the tidal flat east of Chongming Island is northward due to the tide-induced Stokes drift. In the NB, the residual current is weaker, and there is a net water transport toward the SB. The summer residual current during spring tide in the SB, SC, NC, and NP flows seaward, and has a larger amplitude than that in winter due to larger river discharge. The residual current over the tidal flat east of Chongming Island is northward due to tidal pumping and southerly wind. In the NB, the residual current is weaker and almost seaward, and the net transport is from the SB into the NB.

Saltwater intrusion in the Changjiang Estuary is controlled mainly by river discharge and tide, but is also influenced by wind, topography, river basin and estuarine projects, and sea level rise. There is a net landward flow in the NB when river discharge is low during spring tide, resulting in SSO, which is the most striking characteristic of saltwater intrusion in the estuary. Considering wind, sea level rise and the major projects in the river basin and estuary, we simulated and analyzed their impacts on the saltwater intrusion in the Changjiang Estuary using the numerical model. Strong northerly wind enhanced the net water and salt fluxes from the NB into the SB, and enhanced the wind-driven circulation that flowed landward in the NC and seaward in the SC, which favored the saltwater intrusion in the NC. Clearly, the wind is an important dynamic factor of saltwater intrusion in the estuary.

The impact of sea level rise on both the intensity of saltwater intrusion and stratification increase as sea level rises, and have obvious inter-annual changes following different river discharges. The Stokes drift transport is the major mechanism for the up-estuary salt transport in each channel, while the seaward Eulerian transport is of the same order in amplitude. The flux of SSO from the NB into the SB increases due to stronger Stokes transport as sea level rises.

The impact of the deep waterway project is that the saltwater intrusion had been alleviated distinctly in the NC because the dykes of the project blocked off the southward drift of the brackish water plumes under the northerly monsoon and the Coriolis force. The saltwater intrusion in the NP was intensified at the upper section and alleviated at the lower section. In the SP, the saltwater intrusion was intensified as the background salinity increased and the river discharge decreased.

The TGR advanced the timing of saltwater intrusion and slightly increased its intensity during the autumn season, according our modeling results. As the TGR supplemented river discharge during the dry season, the saltwater intrusion was suppressed. The operation of the TGR is basically favorable for reducing the burden of freshwater supply in the highly populated estuarine region.

Hypoxia occurs off the Changjiang mouth because of the presence of excess nutrients, which can cause algal blooms, and of freshwater input, which can intensify stratification. The seasonal variation of pycnocline was consistent with that of hypoxia, and the pycnocline played an important role in preserving hypoxic condition. The residual current speed in the bottom layer was small and favorable for the maintenance of hypoxia during summer.

References

Bianchi, T.S., Allison, M.A., 2009. Large-river delta-front estuaries as natural "recorders" of global environmental change. Proc. Natl. Acad. Sci. U. S. A. 106 (20), 8085–8092.

Blumberg, A.F., Mellor, G.L., 1987. A description of a three-dimensional coastal ocean circulation model. In: Heaps, N.S. (Ed.), Three-Dimensional Coastal Ocean Models. American Geophysical Union, Washington, D.C. pp. 1–16.

Blumberg, A.F., 1994. A primer for ECOM-si. Technical report of HydroQual, Mahwah, NJ.

Chen, C., Zhu, J.R., Ralph, E., Green, S.A., Budd, J.W., Zhang, F.Y., 2001. Prognostic modeling studies of the Keweenaw current in Lake Superior, Part I: formation and evolution. J. Phys. Oceanogr. 31, 379–395.

Chen, C.S., Zhu, J.R., Beardsley, R.C., Franks, P.J.S., 2003. Physical-biological sources for dense algal blooms near the Changjiang. Geophys. Res. Lett. 30 (10), 1515–1518.

Chen, Y., Shi, M.L., Zhao, Y.G. (Eds.), 1989. The Ecological and Environmental Atlas of the three Gorges of the Changjiang. Science Press, Beijing, p. 157 (in Chinese).

Dai, M.H., Guo, X.H., Zhai, W.D., Yuan, L.Y., Wang, B.W., Wang, L.F., Cai, P.H., Tang, T.T., Cai, W.J., 2006. Oxygen depletion in the upper reach of the Pearl River estuary during a winter drought. Mar. Chem. 102, 159–169.

Diaz, R.J., Rosenberg, R., 1995. Marine benthic hypoxia: a review of its ecological effects and the behavioural responses of benthic macrofauna. Oceanogr. Mar. Biol. Annu. Rev. 33, 245–303.

Editorial Board for Marine Atlas, 1992. Ocean Atlas in Huanghai Sea and East China Sea (Hydrology). China Ocean Press, Beijing.

Galperin, B., Kantha, L.H., Hassid, S., Rosati, A., 1988. A quasi-equilibrium turbulent energy model for geophysical flows. J. Atmos. Sci. 45, 55–62.

Hecky, R.E., Kilham, P., 1988. Nutrient limitation of phytoplankton in freshwater and marine environments: a review of recent evidence on the effects of enrichment. Limnol. Oceanogr. 33 (4part2), 796–822.

Kemp, W.M., Boynoton, W.R., Adolf, J.E., Boesch, D.F., Boicourt, W.C., Brush, G., Cornwell, J.C., Fisher, T.R., Glibert, P.M., Hagy, J.D., Harding, L.W., Houde, E.D., Kimmel, D.G., Miller, W.D., Newell, R.I.E., Roman, M.R., Smith, E.M., Stevenson, J.C., 2005. Eutrophication of Chesapeake Bay: historical trends and ecological interactions. Mar. Ecol. Prog. Ser. 303, 1–29.

Li, D.J., Zhang, J., Huang, D.J., Wu, Y., Liang, J., 2002. Oxygen depletion off the Changjiang (Yangtze River) estuary. Sci. China, Seri. D Earth Sci. 45 (12), 1137–1146.

Li, L., Zhu, J.R., Wu, H., Wang, B., 2010. A numerical study on water diversion ratio of the Changjiang (Yangtze) estuary in dry season. Chin. J. Oceanol. Limnol. 28 (3), 700–712.

Li, L., Zhu, J.R., Wu, H., 2012. Impacts of wind stress on saltwater intrusion in the Yangtze estuary. Sci. China Earth Sci. 55 (7), 1178–1192.

Li, L., Zhu, J.R., Wu, H., Guo, Z.G., 2014. Lateral saltwater intrusion in the North Channel of the Changjiang estuary. Estuar. Coasts 37 (1), 36–55.

Mellor, G.L., Yamada, T., 1974. A hierarchy of turbulence closure models for planetary boundary layers. J. Atmos. Sci. 33, 1791–1896.

Mellor, G.L., Yamada, T., 1982. Development of a turbulence closure model for geophysical fluid problem. Rev. Geophys. 20, 851–875.

Pritchard, D.W., 1956. The dynamic structure of a coastal plain estuary. J. Mar. Res. 15, 33–42.

Qiu, C., Zhu, J.R., Gu, Y.L., 2012. Impact of seasonal tide variation on saltwater intrusion in the Changjiang River Estuary. Chin. J. Oceanol. Limnol. 30 (2), 342–351.

Qiu, C., Zhu, J.R., 2013. Influence of seasonal runoff regulation by the Three Gorges reservoir on saltwater intrusion in the Changjiang River estuary. Cont. Shelf Res. 71 (6), 16–26.

Qiu, C., Zhu, J.R., 2015. Assessing the influence of sea level rise on salt transport processes and estuarine circulation in the Changjiang River estuary. J. Coast. Res. 31 (3), 661–670.

Rabalais, N.N., Turner, R.E., Wiseman, W.J., 2001. Hypoxia in the Gulf of Mexico. J. Environ. Qual. 30, 320–329.

Rosenberg, R., Hellman, B., Johansson, B., 1991. Hypoxic tolerance of marine benthic fauna. Mar. Ecol. Prog. Ser. 79, 127–131.

Shen, H.T., Mao, Z.C., Zhu, J.R., 2003. Saltwater Intrusion in the Changjiang Estuary. China Ocean Press, Beijing (in Chinese).

Turner, R.E., Rabalais, N.N., 1994. Coastal eutrophication near the Mississippi River Delta. Nature 368 (6472), 619–621.

Turner, R.E., Rabalais, N.N., Justic, D., 2008. Gulf of Mexico hypoxia: alternate states and a legacy. Environ. Sci. Technol. 42, 2323–2327.

Wang, B.D., 2009. Hydromorphological mechanisms leading to hypoxia off the Changjiang estuary. Mar. Environ. Res. 67, 53–58.

Wei, H., He, Y.C., Li, Q.J., Liu, Z.Y., Wang, H.T., 2007. Summer hypoxia adjacent to the Changjiang estuary. J. Mar. Syst. 67, 292–303.

Wu, H., Zhu, J.R., 2010. Advection scheme with 3rd high-order spatial interpolation at the middle temporal level and its application to saltwater intrusion in the Changjiang estuary. Ocean Model. 33, 33–514.

Wu, H., Zhu, J.R., Chen, B.R., Chen, Y.Z., 2006. Quantitative relationship of runoff and tide to saltwater spilling over from the north branch in the Changjiang estuary: a numerical study. Est. Coast. Shelf Sci. 69, 25–132.

Wu, H., Zhu, J.R., Choi, B.H., 2010. Links between saltwater intrusion and subtidal circulation in the Changjiang estuary: a model-guided study. Cont. Shelf Res. 30, 1891–1905.

Wu, H., Zhu, J.R., Shen, J., Wang, H., 2011. Tidal modulation on the Changjiang River plume in summer. J. Geophys. Res. Atmos. 116 (C8), 192–197.

Xiang, Y.Y., Zhu, J.R., Wu, H., 2009. The impact of the shelf circulations on the saltwater intrusion in the Changjiang estuary in winter. Prog. Nat. Sci. 19 (2), 192–202 (in Chinese).

Zhang, J., 2015. Ecological Continuum from the Changjiang (Yangtze River) Watersheds to the East China Sea Continental Margin. Springer, Switzerland.

Zhang, J., Wu, Y., Zhang, Y.Y., 2015. Plant nutrients and trace elements from the Changjiang Watersheds to the East China Sea. In: Ecological Continuum from the Changjiang (Yangtze River) Watersheds to the East China Sea Continental Margin. Springer, Switzerland, pp. 93–118.

Zhou, M.J., Shen, Z.L., Yu, R.C., 2008. Responses of a coastal phytoplankton community to increased nutrient input from the Changjiang (Yangtze) River. Cont. Shelf Res. 28, 1483–1489.

Zhu, J.R., 2005. Distribution of chlorophyll-a off the Changjiang River and its dynamic cause interpretation. Sci. China Ser. D Earth Sci. 47 (7), 950–956.

Zhu, J.R., Ding, P.X., Zhang, L.Q., Wu, H., Cao, H.J., 2006. Influence of the deep waterway project on the Changjiang estuary. In: The Environment in Asia Pacific Harbours. Springer, The Netherlands, pp. 79–92.

Zhu, J.R., Wu, H., Li, L., Wang, B., 2010. Saltwater intrusion in the Changjiang estuary in the extremely drought hydrological year 2006. J. East China Norm. Univ. Nat. Sci. 4 (1), 1–6 (in Chinese).

Zhu, J.R., Wu, H., Li, L., 2015a. Hydrodynamics of the Changjiang Estuary and adjacent seas. In: Ecological Continuum from the Changjiang (Yangtze River) Watersheds to the East China Sea Continental Margin. Springer International Publishing, Switzerland, pp. 19–45.

Zhu, J.R., Zhu, Z.Y., Lin, J., Wu, H., Zhang, J., 2015b. Distribution of hypoxia and pycnocline off the Changjiang estuary, China. J. Mar. Syst. 154, 28–40.

Zhu, Z.Y., Zhang, J., Wu, Y., Zhang, Y.Y., Lin, J., Liu, S.M., 2011. Hypoxia off the Changjiang (Yangtze River) estuary: Oxygen depletion and organic matter decomposition. Mar. Chem. 125 (1–4), 108–116.

Further Reading

Pawlowicz, R., Beardsley, B., Lentz, S., 2002. Classical tidal harmonic analysis including error estimates in MATLAB using T_TIDE. Comput. Geosci. 28, 929–937.

5

CHANGES IN THE HYDRODYNAMICS OF HANGZHOU BAY DUE TO LAND RECLAMATION IN THE PAST 60 YEARS

Li Li[*,†], Taoyan Ye[*], Xiao Hua Wang[†,‡], Zhiguo He[*,†], Ming Shao[*]

*Ocean College, Zhejiang University, Zhoushan, China †State Key Laboratory of Satellite Ocean Environment Dynamics, Second Institute of Oceanography, Hangzhou, China ‡The Sino-Australian Research Centre for Coastal Management, The University of New South Wales, Canberra, ACT, Australia

CHAPTER OUTLINE
1 Introduction   78
2 Urbanization   78
3 Changes of Physical Environment   82
  3.1 Field Data Analysis   82
  3.2 Numerical Study of Changes of Tides in the Bay   82
4 Discussion   90
  4.1 The Impact of Coastal Engineering on the Tidal Bore   90
  4.2 The Stability of the Bay Under the Impact of Coastal Engineering   90
5 Conclusions   91
Acknowledgments   91
References   91

Sediment Dynamics of Chinese Muddy Coasts and Estuaries. https://doi.org/10.1016/B978-0-12-811977-8.00005-4

1 Introduction

As a macro-tidal estuary, Hangzhou Bay (HZB, Fig. 1A) is well known for its magnificent tidal bore. The bay, located on the coast of the East China Sea, is the estuary of the Qiantang River (QR). The Hangzhou Bay-Qiangtang River system and its major tributaries serve as important cultural, economic, and ecological resources locally and nationally in China.

Hangzhou Bay is a funnel-shaped semidiurnal macro-tidal bay. The width of the bay decreases from 100 km at the bay mouth to about 20 km at the Station Ganpu, over a distance of only 95 km. Tidal range at the mouth (2.2 m at Chuanshan, marked as CS in Fig. 1A) is smaller in the south than that in the north at Luchaogang (LCG, 3.2 m). Tidal range increases westward from about 3.0 m at the mouth to about 5.7 m at Ganpu (GP), while the difference between the south and the north decreases gradually from about 1.0 m to <0.5 m. Flood currents converge off the coast of Jinshan (JS) and are enhanced strongly due to the restriction of the geometry of the bay. Flood current velocity increases westward from about 1.5 ms^{-1} at the mouth to about 1.8 ms^{-1} off the coast of Jinshan, due to the influence of the Changjiang Estuary. Generally, the flood velocity is larger than the ebb velocity in the northern bay. The ebb velocity dominates slightly in the southeastern bay, while the flood velocity is close to the ebb velocity in the central bay (Xie et al., 2013).

Affected by both the sediment form the Changjiang Estuary and sediment re-suspension, HZB is one of the most turbid estuaries in the world. Surrounded by one of the most urbanized areas in China, continuous human intervention, for example, reclamation and navigation channel building, in the last several decades to facilitate commercial navigation, flood-risk mitigation, hydropower, and water supply have altered the hydrologic, hydraulic, and geomorphic processes in the bay. Reclamation and other coastal engineering practices have substantially impacted the flow and sediment transport regimes in the Hangzhou Bay-Qiantang River system.

2 Urbanization

With the development of marine economy and coastal engineering around HZB, the demand for land, especially coastal land, is increasingly higher. Reclaiming new land from the sea is one potential way to meet these demands (Niu and Yu, 2008). Details of the changes in the coastline around Hangzhou Bay over a

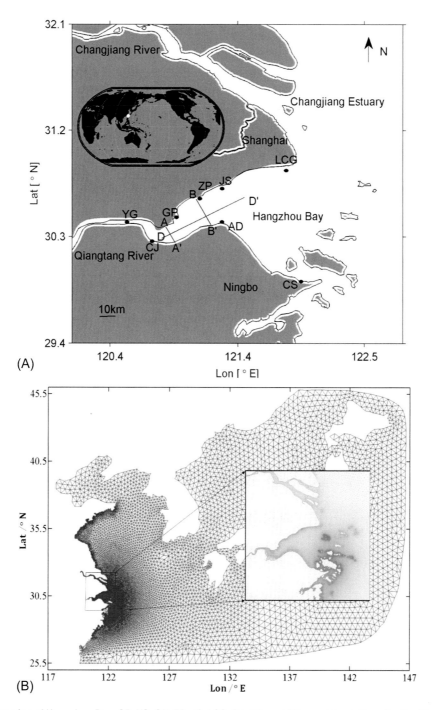

Fig. 1 (A) Location of Hangzhou Bay. QR, YG, GP, ZP, JS, LCG, CJ, AD, and CS are abbreviations for the Qiantang River, Yanguan, Ganpu, Zhapu, Jinshan, Luchangang, Caoejiang, Andong, and Chuanshan. (B) Numerical model grids. The small domain is the finer grid nested into the large domain.

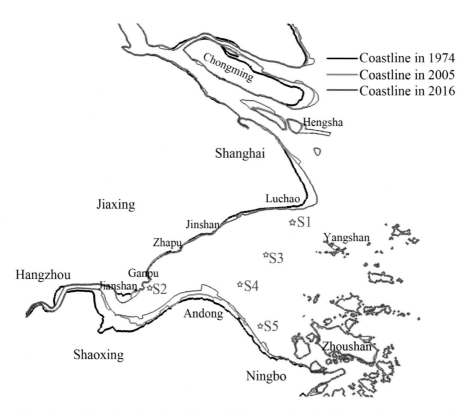

Fig. 2 Coastline change of Hangzhou Bay from 1974 to 2016. Data are from LandSat satellite images. The S1 to S5 are field stations for observation data.

period of nearly 60 years can be obtained from remote sensing images of LandSat (Fig. 2). The land reclamation around HZB was mostly near its south bank. As shown by Zhang et al. (2005), about 694.4 km^2 land was reclaimed around the bay over a period of nearly 30 years from 1986. By 2012, about 94% of the coastline around HZB was artificial coastline (Kemp et al., 2016).

According to the results of remote sensing data analysis (Table 1), most of the reclamation happened at the south bank near Andong (AD) (Fig. 2) during 1962–2005. From 2005 to 2015, some of the reclamation was conducted at the north bank near the Station Yanguan (YG). The width of the Ganpu Section (Section AA′) was reduced by 760 m by 2005, which was 3% of the width (23.1 km) in 1962, and it further reduced by about 3233 m by 2015, which was 14% of the width in 1962. The total reduction of the Zhapu Section (Section BB′) from 1962 to 2015 was 4866 m, which was 14.5% of the width (33.7 km) in 1962. These data are shown in Table 2.

Table 1 Sources of remote sensing images

Time	Data type	Band	Spatial resolution (m)
1973-03-19	LM	7	80
2005-02-27	ETM+	7	15
2015-03-12	ETM+	8	15

The south bank is apt to siltation near Andong due to different siltation processes of sediments from the Changjiang Estuary and the Qiantiang River (Diop and Scheren, 2016), which is favorable for reclamation and aquaculture. The development of aquaculture was an important reason for the change at the south bank (Zhang et al., 2005). The north bank is inclined to erosion, which is favorable for ports. Hence, the changes at the north bank were mostly due to coastal engineering.

These urbanization activities have attracted researchers both in China and overseas. Past research indicated that a single, small area of reclamation activity only had negligible impact on astronomical tides, but the accumulated effect of reclamation can affect tides, with respect to tidal prism and water exchange (Zeng et al., 2011). For example, the impacts of small areas of land reclamation on astronomical tides were small compared with the original magnitudes, but the impacts on local shallow water tides were no longer negligible (Gao et al., 2014; Li et al., 2014; Li et al., 2012; Lin et al., 2015); the constructions of bridges caused the migration of sediment deposition (Ma et al., 2013; Qiao et al., 2011). Therefore, some coastal engineering projects, for example, reclaimed tidal flats, were restored or planned to create artificial tidal flats to mitigate lost ones (Lee et al., 1998; Sohma et al., 2009).

Table 2 Statistics of total reclamation areas

Time period	Increase of reclamation area (km^2)	Decrease of AA' width (m)	Ratio of width to 1962	Decrease of BB' width (m)	Ratio of width to 1962
1962–2005	564.1	761.6	3%	1611.9	4.8%
2005–2015	174.6	3233.4	14%	3253.9	9.7%

Note: the reclamation areas are from Song et al. (2007). The reduction of cross-section width was calculated from satellite data.

3 Changes of Physical Environment

3.1 Field Data Analysis

Previous studies showed that the tidal range has increased by 0.57–1.05 m along HZB and the siltation of the Qiangtang River has happened since the 1990s (You et al., 2010). According to field data analysis (Kidwai et al., 2016), the average high water level and low water level, before (1955–1957) and after (1988–1991) a reclamation period, were increased. As shown in Table 3, the high-water level at Ganpu and Zhapu increased by 0.32 m and 0.21 m, respectively. The low water level had just a slight increase. Hence, only the results during spring tides are shown in the following sections.

3.2 Numerical Study of Changes of Tides in the Bay

3.2.1 Model Setup and Validation

The finite volume coastal ocean model (FVCOM) (Chen et al., 2003) was chosen to simulate the hydrodynamics of HZB. FVCOM is a three-dimensional hydrodynamic model, which uses an un-structured grid. The continuity equation is

$$\frac{\partial \zeta}{\partial t} + \frac{\partial (Du)}{\partial x} + \frac{\partial (Dv)}{\partial y} + \frac{\partial w}{\partial \sigma} = 0 \qquad (1)$$

where x, y and σ are the eastward, northward, and vertical coordinates, respectively; and u, v and w are the corresponding velocity components; and ζ is the height of the free surface, D is the total water-column depth. The model employs the Mellor-Yamada level-2.5 turbulence closure scheme (Mellor and Yamada, 1982)

Table 3 Observed tidal elevation before and after reclamation

Station	High water (m)			Low water (m)		
	Before	*After*	*Different*	*Before*	*After*	*Different*
Ganpu	7.19	7.51	0.32	1.83	1.83	0.00
Zhapu	6.13	6.94	0.21	2.22	2.23	0.01

Data from Kidwai, S., Fanning, P., Ahmed, W., Tabrez, M., Zhang, J., Khan, M.W., 2016. Practicality of marine protected areas – can there be solutions for the river Indus delta? Estuar. Coast. Shelf Sci. 183 (Part B), 349–359. https://doi.org/10.1016/j.ecss.2016.09.016.

for vertical mixing, and uses the Smagorinsky scheme (Smagorinsky, 1963) for horizontal mixing.

The unstructured triangular grids of the model domain consist of 62,749 elements and 32,829 nodes (Fig. 1B), with the cell sizes of the elements ranging from 100 m near islands to 50 km at the ocean open boundary. Ten uniform vertical layers are specified in the water column using the σ-coordinate system. Four stations that routinely collect elevation data are YG, GP, ZP, and JS (Fig. 1A). Another five stations of S1 to S5 (Fig. 2) are for collecting current data. Also indicated in Fig. 1A are the sections AA', BB', and DD' for analysis of vertical current profiles.

Two main rivers, the Changjiang River (CR) and Qiantang River (QR), discharge into this region. According to records of annual-averaged flux, 60,000 m^3s^{-1} is set for the CR while 864 m^3s^{-1} for the QR in the model. To focus on the effects of the reclamation, the wind is ignored in this study. At the open boundary, four diurnal components (K_1, Q_1, P_1, Q_1) and four semidiurnal components (M_2, K_2, S_2, N_2), which are specified by using the TPXO7.2 global model of ocean tides (Egbert, 1997), are used to calculate water level. Of all the tidal components, M_2 plays the most important role in the tidal forcing. As our focus is the tidal dynamics of HZB, both salinity and temperature are initialized with constants. Three model runs for 1974, 2005, and 2016 are designed to examine the accumulated effects of reclamation on tidal dynamics. The model run for 2005 is the reference, as the field data for model calibration were available for this year.

After being calibrated by field data of tidal elevation and currents (Fig. 3), the model reproduces the hydrodynamics in the bay. The normalized root mean square errors for tidal elevation at the five stations are between 0.2 and 0.5. The skill score (Zhu et al., 2016) calculated by Eq. (2) were all larger than 0.9. The deviations for the vertically averaged current speed and phase are between 0.2 ms^{-1} and 0.5 ms^{-1}, and between 26 degrees and 46 degrees, respectively.

$$SS = 1 - \frac{\sum_{i=1}^{N}(m_i - o_i)^2}{\sum_{i=1}^{N}(o_i - \overline{o})^2} \qquad (2)$$

where m_i and o_i are the predicted and observed values of SSC at time t, which have mean values and, and standard deviations S_m and S_o, respectively.

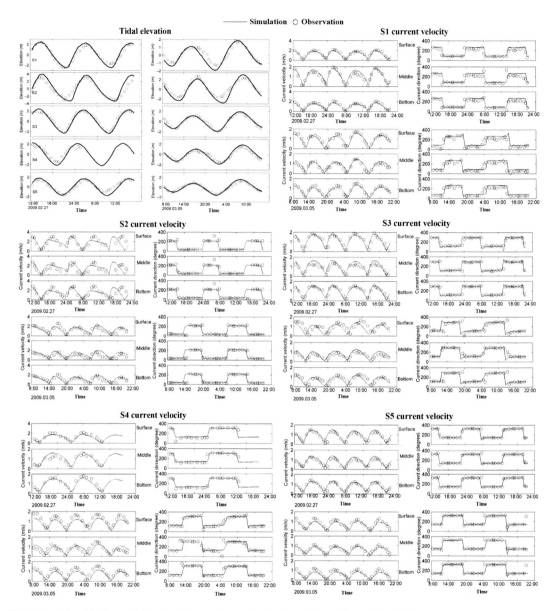

Fig. 3 Model validation of tidal elevation and current velocity at the stations S1 to S5.

3.2.2   Changes of Tidal Range

According to the model results, the maximum tidal range upstream of the Ganpu section increased >1 m from 1974 to 2005 (Fig. 4A), which is in accordance with the observed data shown by Kidwai et al. (2016). The largest increase in this period can reach 2 m near the Station Jianshan. This is due to the reclamation

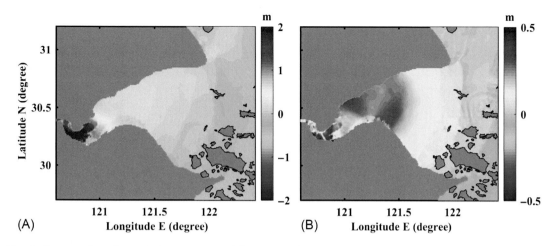

Fig. 4 Changes of maximum tidal range during (A) 1974–2005 (year 2005 minus year 1974) and (B) 2005–2016 (year 2016 minus year 2005).

between 1974 and 2005, which mostly happened around the head of the bay and directly reduced the tidal prism there. The water level near the mouth of the bay was only affected slightly.

From 2005 to 2016, the tidal range had another increase of about 0.3 m upstream of the Zhapu section (Fig. 4B). The largest increase of the maximum tidal range from 2005 to 2016 appeared near the Zhapu section. In this stage, the reclamation was mainly around Andong (AD in Fig. 1A), opposite to the Station Zhapu, which narrowed the width of the Zhapu cross-section.

Reclamation reduced the tidal prism of the bay, and consequently amplified the high-water level inside the bay. The reduction of the width near the Ganpu Section amplified the shoaling effect, and consequently increased the tidal level at the Ganpu Section. The decrease of the width at the Section Jianshan in 2016 (Fig. 1A) amplified the tidal choking effect and then dampened the tidal amplitude upstream of the Jianshan Section. The tidal level at the Yanguan Section was only slightly increased due to the combined effect of increased shoaling effect and increased tidal choking effect (Fig. 4B).

3.2.3 Changes of Currents

The vertically averaged tidal ellipses of M_2 tide at Jinshan, Zhapu, Ganpu, and Caoejiang are displayed in Fig. 5. At the Station Jinshan and the Station Zhapu, the magnitudes and directions of tidal ellipses in the 3 years (1974, 2005, and 2016) remained almost the same. At the Station Ganpu, the major axis of the M_2 tide were reduced from about 1.0 ms^{-1} in 1974 to 0.8 ms^{-1} in 2016. At the Station

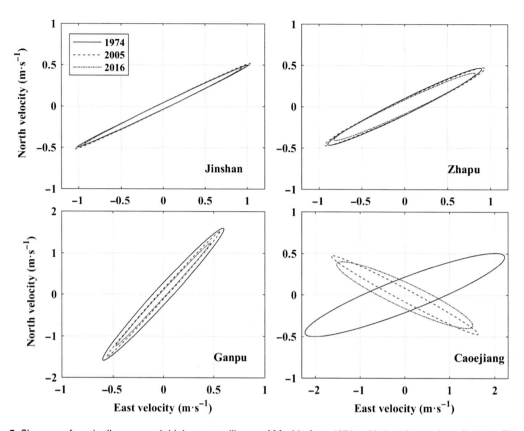

Fig. 5 Changes of vertically averaged tidal current ellipses of M_2 tide from 1974 to 2016 at the stations Jinshan, Zhapu, Ganpu, and Caoejiang in Hangzhou Bay.

Caoejiang, the directions and magnitudes of tidal ellipses had the largest variation among all the four stations. The direction changed from almost southwest-northeast direction to the northwest-southeast direction. The magnitude decreased from 2.2 ms^{-1} in 1974 to about 1.5 ms^{-1} in 2005 and 2016.

3.2.4 Changes of Main Tidal Channel

Two cross-sections of AA′ (near the Station Ganpu) (Fig. 6) and DD′ (from the head to the mouth of the bay) (Fig. 7) are selected to display the vertical profiles of flooding and ebbing currents and the changes of tidal channel during a spring tidal cycle in the bay. Tides flooded into the bay along the north bank (Fig. 6A–C) and ebbed along the south bank (Fig. 6A′–C′) at the Ganpu Section. There were three tidal channels in 1974 and 2005, but only two of them remained in 2016. The main flooding/ebbing tidal

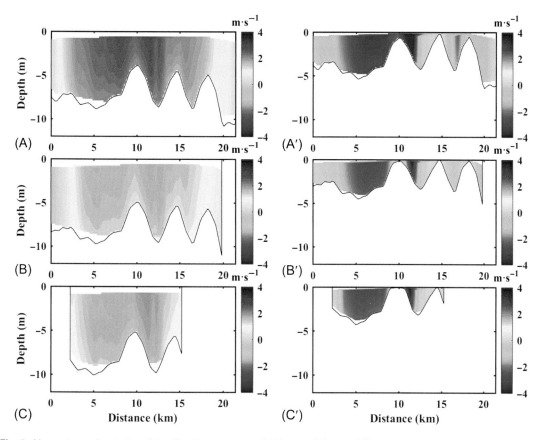

Fig. 6 Along-channel velocity of the flooding currents of (A) 1974, (B) 2005, (C) 2016, and the ebbing currents of (A') 1974, (B') 2005, (C') 2016 during the spring tides at the Section AA'. The direction from left to right is from A to A'. Positive values are for landward currents.

channel moved slightly from the center of the section (in 1974) to near the north bank in 2005 after reclamation. In 2016, the tidal channel was similar, but narrower than that in 2005, again due to the changes in geometry after reclamation. The radius of the curvature was decreased after reclamation, and the centrifugal force was increased consequently. Hence, the tidal channel moved toward the north bank due to increased centrifugal force.

At the Ganpu Section (Section AA'), the peak flooding current speed decreased from $2.9\,\text{ms}^{-1}$ in 1974 to $2.5\,\text{ms}^{-1}$ in 2005, and slightly increased to $2.6\,\text{ms}^{-1}$ in 2016. During ebbing tides, the peak current speeds decreased from 1974 ($3.4\,\text{ms}^{-1}$) to 2005 ($2.9\,\text{ms}^{-1}$), and then increased slightly in 2016 ($3.2\,\text{ms}^{-1}$). This was because of the reduced bottom friction caused by decreased seabed area and increased tidal level in the bay.

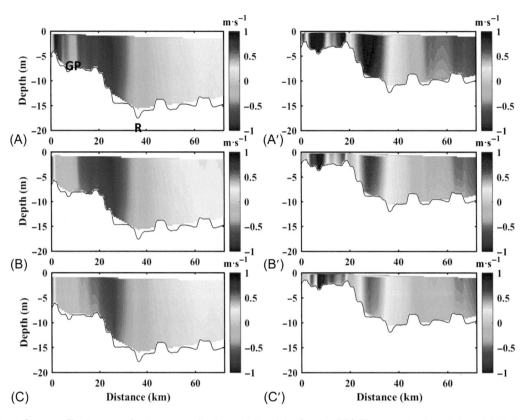

Fig. 7 Same as Fig. 6, except for the perpendicular velocity of the Section DD′. The direction from left to right is from D to D′. *Positive* values indicate the northward currents.

Tides propagate sinuously into the bay through the Section DD′ from south to north at the mouth (right-hand end of the Section DD′, Fig. 7), and then from north to south at the middle part. The reverse of the directions happened at the shallow area of the bay (indicated by "R" in Fig. 7). The main tidal channel, where the peak current speed occurred, became narrowed and moved seaward from 1974 to 2016. In the middle part of the bay (between "GP" and "R" in Fig. 7), the transverse current speed, at the core of the maximum velocity zone during the spring tide, decreased by about 0.4 ms^{-1} during the peak current from 1974 to 2016.

3.2.5 Changes of Residual Currents

Residual currents are important for sediment and nutrient transports. The residual current velocity around the Zhoushan Islands is calculated over 30 tidal cycles using:

$$U = \frac{1}{nT} \int_{t_0}^{t_0+nt} u(x, y, t)dt \qquad (3)$$

$$V = \frac{1}{nT} \int_{t_0}^{t_0+nt} v(x, y, t)dt$$

where t_0 is the start time, T is the tidal period. n is the number of tidal cycles. $u(x, y, t)$ and $v(x, y, t)$ are the current velocities in the x and y directions, respectively; U and V are the Eulerian-averaged current velocities in the x and y directions, respectively.

According to the reference model (the model for 2005), the vertically-averaged peak residual current speed of about $0.8 \, \mathrm{ms}^{-1}$ appeared near Andong in 1974, but this peak area became smaller in 2005 and 2016. The vertically-averaged residual currents were all seaward in the entire bay in the 3 years due to the Qiangtang River discharge, but with decreased magnitudes of $<0.2 \, \mathrm{ms}^{-1}$ from 1974 to 2016. The largest decrease happened near the Ganpu Section.

Upstream of the Station Ganpu, the seaward residual currents became weaker (decrease by about $0.2 \, \mathrm{ms}^{-1}$) in 2005 after reclamation, and then slightly decreased in 2016 (Fig. 8). Upstream of the Jianshan Station, the seaward residual currents became stronger in 2016 after reclamation. Residual currents remained almost the same in other areas.

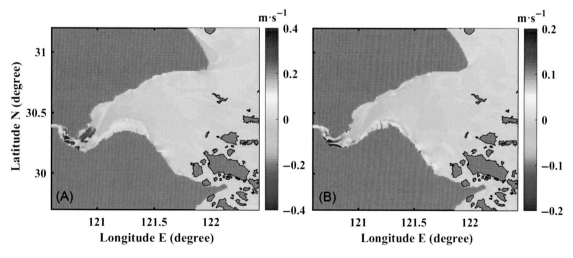

Fig. 8 Differences of vertically-averaged residual currents: (A) during 1974–2005 (year 2005 minus year 1974) and (B) during 2005–2016 (year 2016 minus year 2005) in Hangzhou Bay.

4 Discussion

4.1 The Impact of Coastal Engineering on the Tidal Bore

Decreased width at the head of the Hangzhou Bay increased the shoaling effect in the bay. The tidal amplitude at the Station Yanguan was amplified due to the increased shoaling effect. Reclamation at the Jianshan Section increased the tidal choking effect and consequently dampened the tidal amplitude at the Station Yanguan. The impact of the shoaling effect overtook that of the tidal choking effect, and the tidal amplitude at the Station Yanguan was slightly increased. Hence, if the width of the Jianshan Section is further reduced, the level of the tidal bore near the Station Yanguan might be dampened, and consequently, tidal bore tourism would be negatively impacted.

The increase of tidal choking effect would also cause the slightly decrease/increase of the tidal currents seaward/landward from the Station Jianshan. These changes of tidal currents would tend to induce sediment siltation downstream and bottom erosion upstream from the Station Jianshan, respectively. Hence, the tidal bore would be further reduced because of the reduction of the shoaling effect due to the bottom erosion in the future.

4.2 The Stability of the Bay Under the Impact of Coastal Engineering

Bottom erosion would be reduced and less sediment would be eroded downstream of the Ganpu Section, due to the reduced tidal currents. In the long run, this is not a stable situation for the maintenance of the coastal engineering, which prefers slight erosion to maintain the depth of the navigational channel.

The main tidal channel migrated northward and narrowed after 2016 due to the increased centrifugal force. This migration of tidal channel would impact the stability of the navigational channel and the coastal facilities.

The seaward residual currents upstream/downstream from the Station Jianshan were amplified/dampened respectively, from 1974 to 2016. Hence, sediment tends to be accumulated between the Section Jianshan and the Section Ganpu, which is not favorable to navigation and coastal stability.

Therefore, the accumulated impact of coastal engineering of the Hangzhou Bay changed the hydrodynamics in the bay, and consequently caused an unstable situation in the navigational channel and the coastal facilities in/around the bay.

5   Conclusions

The accumulated urbanization around HZB in the period of near 60 years narrowed the width of the bay and changed the hydrodynamics of the bay. In 2015, the reclaimed widths of the Section Ganpu and the Section Zhapu were all >3 km, being 14% and 9.7% of that in 1974, respectively.

The maximum tidal range increased by >2 m upstream from the Ganpu Section from 1974 to 2016 due to the combined effects of the increased shoaling effect and the increased tidal choking effect. The tidal currents decreased due to the reduced tidal prism at the Stations Ganpu and Caoejiang. The tidal current directions changed due to the controlling effect of the narrowed coastline. At Section Ganpu, the main tidal channel migrated northward and narrowed after 2016 due to increased centrifugal force. The core velocity of the currents in the main tidal channel reduced during the spring tides from 1974 to 2016. The seaward residual currents upstream/downstream from the Station Jianshan were amplified/dampened by about $0.2\,\mathrm{ms}^{-1}$, respectively, from 1974 to 2016.

Changes in the coastline would impact the tidal bore and probably cause an unstable situation for the estuary. Hence, the design and planning of the navigational channel and the coastal facilities in estuaries should take into careful consideration the correlation between the planned projects and the hydrodynamics/sediment dynamics in the estuaries.

Acknowledgments

This research was supported by the National Natural Science Foundation of China (Grant No. 41606103), the Natural Science Foundation of Zhejiang Province of China (Grant No. Q16D060002, LR16E090001), the State Key Laboratory of Satellite Ocean Environment Dynamics at the Second Institute of Oceanography, State Oceanic Administration (Grant No. SOED1512), and the National Key Research and Development Program of China (Grant No. 2017YFC1405101).

References

Chen, C., Liu, H., Beardsley, R.C., 2003. An unstructured grid, finite-volume, three-dimensional, primitive equation ocean model: application to coastal ocean and estuaries. J. Atmos. Ocean. Technol. 20 (1), 159–186. https://doi.org/10.1175/1520-0426(2003)020<0159:AUGFVT>2.0.CO;2.

Diop, S., Scheren, P.A., 2016. Sustainable oceans and coasts: lessons learnt from eastern and western Africa. Estuar. Coast. Shelf Sci. 183 (Part B), 327–339. https://doi.org/10.1016/j.ecss.2016.03.032.

Egbert, G.D., 1997. Tidal data inversion: interpolation and inference. Prog. Oceanogr. 40 (1–4), 53–80. https://doi.org/10.1016/S0079-6611(97)00023-2.

Gao, G., Wang, X.H., Bao, X.W., 2014. Land reclamation and its impact on tidal dynamics in Jiaozhou Bay, Qingdao, China. Estuar. Coast. Shelf Sci. 151, 285–294. https://doi.org/10.1016/j.ecss.2014.07.017.

Kemp, G.P., Day, J.W., Rogers, J.D., Giosan, L., Peyronnin, N., 2016. Enhancing mud supply from the lower Missouri River to the Mississippi River Delta USA: dam bypassing and coastal restoration. Estuar. Coast. Shelf Sci. 183 (Part B), 304–313. https://doi.org/10.1016/j.ecss.2016.07.008.

Kidwai, S., Fanning, P., Ahmed, W., Tabrez, M., Zhang, J., Khan, M.W., 2016. Practicality of marine protected areas – can there be solutions for the river Indus delta? Estuar. Coast. Shelf Sci. 183 (Part B), 349–359. https://doi.org/10.1016/j.ecss.2016.09.016.

Lee, J.G., Nishijima, W., Mukai, T., Takimoto, K., Seiki, T., Hiraoka, K., Okada, M., 1998. Factors to determine the functions and structures in natural and constructed tidal flats. Water Res. 32 (9), 2601–2606. https://doi.org/10.1016/S0043-1354(98)00013-X.

Li, L., Wang, X.H., Andutta, F., Williams, D., 2014. Effects of mangroves and tidal flats on suspended-sediment dynamics: observational and numerical study of Darwin Harbour, Australia. J. Geophys. Res. Oceans 119 (9), 5854–5873.

Li, L., Wang, X.H., Williams, D., Sidhu, H., Song, D., 2012. Numerical study of the effects of mangrove areas and tidal flats on tides: a case study of Darwin Harbour, Australia. J. Geophys. Res. Oceans 117(C6). https://doi.org/10.1029/2011jc007494.

Lin, I., Liu, Z., Xie, L., Gao, H., Cai, Z., Chen, Z., Zhao, J., 2015. Dynamics governing the response of tidal current along the mouth of Jiaozhou Bay to land reclamation. J. Geophys. Res. Oceans 120 (4), 2958–2972. https://doi.org/10.1002/2014JC010434.

Ma, G., Shi, F., Liu, S., Qi, D., 2013. Migration of sediment deposition due to the construction of large-scale structures in Changjiang estuary. Appl. Ocean Res. 43, 148–156. https://doi.org/10.1016/j.apor.2013.09.002.

Mellor, G.L., Yamada, T., 1982. Development of a turbulent closure model for geophysical fluid problems. Reviews of Geophysics 20 (4), 851–875. https://doi.org/10.1029/RG020i004p00851.

Niu, X., Yu, X., 2008. A practical model for the decay of random waves on muddy beaches. J. Hydrodyn. Ser. B 20 (3), 288–292. https://doi.org/10.1016/S1001-6058(08)60059-1.

Qiao, S., Pan, D., He, X., Cui, Q., 2011. Numerical study of the influence of Donghai bridge on sediment transport in the Mouth of Hangzhou Bay. Procedia Environ. Sci. 10 (Part A(0)), 408–413. https://doi.org/10.1016/j.proenv.2011.09.067.

Smagorinsky, J., 1963. General circulation experiments with the primitive equations. Monthly Weather Review 91 (3), 99–164.

Sohma, A., Sekiguchi, Y., Nakata, K., 2009. Application of an ecosystem model for the environmental assessment of the reclamation and mitigation plans for seagrass beds in Atsumi Bay. Estuar. Coast. Shelf Sci. 83 (2), 133–147. https://doi.org/10.1016/j.ecss.2007.11.030.

Song, L., Wang, X., Xiang, W., Zhou, Q., 2007. Analysis of remote sensing dynamic monitoring for tidal zone resources of Hangzhou Bay. Zhejiang Hydrotechnics 1, 11–17, 1008-701X (2007) 01-0011-07.

Xie, D.-F., Gao, S., Wang, Z.-B., Pan, C.-H., 2013. Numerical modeling of tidal currents, sediment transport and morphological evolution in Hangzhou Bay, China. Int. J. Sediment Res. 28 (3), 316–328. https://doi.org/10.1016/S1001-6279(13)60042-6.

You, A., Han, Z., He, R., 2010. Characteristics and effecting factors of the tidal level in the Qiantangjiang River Estuary under changing environment. J. Mar. Sci. 28 (1), 18–25. https://doi.org/1001-909X(2010)01-0018-08.

Zeng, X., Guan, W., Pan, C., 2011. Cumulative influence of long term reclamation on hydrodynamics in the Xiangshangang Bay. J. Mar. Sci. 29 (1), 73–83. https://doi.org/1001-909X(2011)01-0073-11.

Zhang, H., Guo, Y., Huang, W., Zhou, C., 2005. A remote sensing investigation of inning and silting in Hangzhou Bay since 1986. Remote Sens. Land Resour. 2, 50–54. https://doi.org/1001-070X(2005)02-0050-05.

Zhu, L., He, Q., Shen, J., Wang, Y., 2016. The influence of human activities on morphodynamics and alteration of sediment source and sink in the Changjiang estuary. Geomorphology 273, 52–62. https://doi.org/10.1016/j.geomorph.2016.07.025.

6

MARINE ENVIRONMENTAL STATUS AND BLUE BAY REMEDIATION IN XIAMEN

Keliang Chen*, Senyang Xie[†,‡], Hongzhe Chen*

*Third Institute of Oceanography, State Oceanic Administration, Xiamen, China;
[†]School of Physical, Environmental, and Mathematical Sciences, University of New South Wales, Canberra, ACT, Australia; [‡]The Sino-Australian Research Centre for Coastal Management, University of New South Wales, Canberra, ACT, Australia

CHAPTER OUTLINE

1 **Climatology, Hydrology, and Geology 95**
2 **Biodiversity and Ecological Disasters 100**
3 **Human Interventions: Reclamation and Its Environmental Impacts 104**
    3.1 Extensive Reclamation Works in Xiamen Bay From the Year 1955 104
    3.2 Environmental Impacts 105
4 **Modeling Approach 108**
    4.1 Introduction of the Environmental Fluid Dynamics Code (EFDC): A Powerful Tool for Coastal Environmental Management 108
    4.2 Application of EFDC to Study Suspended Sediment Concentration in Xiamen Bay 109
5 **Recommendations and Future Steps 117**
References 120
Further Reading 122

1 Climatology, Hydrology, and Geology

Xiamen Bay is located on the west coast of the Taiwan Strait, adjacent to the city of Xiamen (formerly romanized as Amoy) in Fujian Province of China. In geography, the outer boundary of

Sediment Dynamics of Chinese Muddy Coasts and Estuaries. https://doi.org/10.1016/B978-0-12-811977-8.00006-6

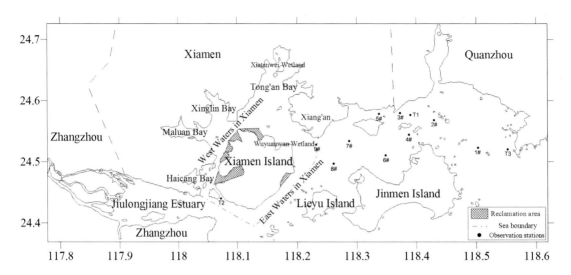

Fig. 1 Map of Xiamen and Jinmen.

Xiamen Bay is from Cape Weitou of Jinjiang County, Quanzhou, to Cape Zhenghai of Longhai County, Zhangzhou; and it also involves the Kinmen (based on the postal romanization, with the Chinese Phonetic Alphabet form of Jinmen) Waters under the jurisdiction of the Taiwan authorities (Fig. 1). However, in most cases Xiamen Bay is referred to the waters under the jurisdiction of Xiamen City for research and management. Xiamen Bay in this chapter is also referred to the waters under the jurisdiction of Xiamen City. The waters in Xiamen Bay are traditionally divided into four main areas, namely, the Jiulong Estuary, West Waters in Xiamen, East Waters in Xiamen, and Tong'an Bay. There are Xiamen Island and several islets in Xiamen Bay. As the most important part, the West Waters consist of several semienclosed bays (such as Xinglin Bay, Haicang Bay, Maluan Bay) and important shipping lanes (such as Xiamen-Gulangyu Waterway, Songyu Waterway). There are main berths in Xiamen Outer Port, Xiamen West Port, and Free Trade Zone, which are close to a tourist attraction (Gulangyu Islet-Zhongshan Road scenic area). As a result, the passing ships and tourist spots surrounding this area become various pollution sources.

Xiamen City, which is surrounded by Zhangzhou City, Quanzhou City, and Kinmen County, is a tourist city, famous for its attractive seascape. It is also a special economic zone in China. Xiamen Port is the most important deep-water port in the southeast coast of China, and the ninth largest port in China. According to Lloyd's List 2016, Xiamen Port is the sixteenth largest international container port (TOP 100 Container Ports in 2016, 2016). At the end of 2015, the mariculture area of Xiamen Bay has dropped to $31.18\,km^2$ (2016 Yearbook of Xiamen Special Economic Zone, 2016). In addition, there are many bathing beaches in the south and east of Xiamen Island, while several wetland parks are located near the West Waters and Tong'an Bay. Therefore, a good marine environmental quality has a vital role in Xiamen's economy for tourism and its related industries.

Xiamen City is one of the initial four special economic zones and one of the free trade zones in China. According to the 2016 Yearbook of Xiamen Special Economic Zone (2016 Yearbook of Xiamen Special Economic Zone, 2016), Xiamen City is a coastal city with a total land area of $1699.39\,km^2$, and its sea area is of $390\,km^2$, along with a coastal line of $234\,km$; the land area of Xiamen Island is of $157.76\,km^2$. The city enjoys US\$14,000 per capita GDP per year. At the end of 2015, the resident population of Xiamen City reached 3.86 million, 2.5 million of which lived on Xiamen Island. The population density of Xiamen Island is over $12,700$ persons/km^2. The population growth has caused great environmental pressure on marine environment. Since the city was established in 1933, Xiamen's growth has been constrained by limited land area, consequent traffic inconvenience and coastal erosion. Land reclamation and construction of seawalls (such as the Maluan Bay Seawall and the Gaoji Seawall) have been carried out by the municipal managers since the 1950s. These activities have changed the topography and coastline of Xiamen Bay greatly. In recent years, as part of Xiamen's ecological restoration process, some wetlands and beaches have been restored, and several seawalls have been demolished.

The Jiulong River is the second largest river in Fujian Province, with an average annual runoff of $14,126$ million m^3 according to the government open data for 2006–2011 (Hydrological Information Network of Fujian Province, n.d.). Along with the surface deforestation and the erosion of mountain earthwork, the average annual sediment amount of the Jiulong River increased to around 2.23 Mt. after 2000, with the maximum annual sediment yield over 6.47 Mt. (Xu and Li, 2003). Due to the small runoff, the waters of the Jiulong Estuary are highly affected by tides. The survey data during 1985–2009 (Bao, 2011) indicated that, with river erosion and siltation, the modern subaqueous delta in the Jiulong Estuary is stretching forward toward Xiamen Harbor. The scale of tidal delta in the narrow channel lying adjacent to the south of Kinmen Island is also expanding, and the shape has changed from a fan into a tongue. Additionally, there are many short and small streams entering Xiamen Bay, such as the Hou'xi River into Xinglin

Bay, the Shenqing'xi River and Guoyun'xi River into Maluan Bay, the Xi'xi River, Dong'xi River, Guanxun'xi River, and Litou'xi River into Tong'an Bay; all these bays are part of Xiamen Bay. Of these rivers, the largest annual flow of the Xi'xi River is estimated to be about 370 million m^3.

The tides in Xiamen Bay are semidiurnal. The maximum tidal range is 6.00 m, and the minimum is 0.99 m, with an average of 3.00 m (Liu et al., 1984). Rapid water flows caused by numerous islets and strong tides lead to numerous tidal inlets in the floor of Xiamen Bay, with alternating ridge and trough. Moreover, there are landform types like tidal sand ridge, tidal delta, modern sub-aqueous delta, underwater slope, and shelf plains. Though the difference is large, the water depth in the bay is generally <20 m. For example, the water depths over about 64 km^2 of submarine reefs are between 20 m and 32 m, while those over the tidal sand ridges are from 10 m to 22 m. The ridge lines are on the south side of sand ridges mostly, with NE-SW trending and parallel/banded distribution. The north slopes are gentle, and the south slopes are steep. Because the seabed near the main waterway has been dredged regularly, the water depth there is large; for example, the depth near the Dongdu Channel is over 30 m.

The topography was formed mainly by tectonic movement since the Late Triassic period. The rise of sea level, the supply, transport and deposition of sediment, and hydrodynamic forces all contributed to the strata formation. The lithology is composed of granite and igneous rocks of the Yanshan period, covered by a 30-m thick Holocene sediment (Hong and Chen, 2003), including a 4-m thick in surface sediment in most situation (Cai et al., 1993). There are no confined aquifers (Wang et al., 2015). In terms of sediment type, the main component of surface sediments in Xiamen Bay is fine clayey silt. Coarse sands could only be found in the sediments near the Jiulong Estuary, and sediment particles become smaller along the stream, but sediment particles of the East Waters are complex, with main composition of coarse gravel and coarse sand (Zuo et al., 2011). Compared with the data of the 1980s, the sediment particles in the West Waters and Tong'an Bay have become coarser gradually (Fang, 2008). Most of the sediment minerals are terrestrial. Roughly half of them are silicates (Hong and Chen, 2003), lacking phosphorus mineral (Li et al., 1999). There are also magnetite, hematite, epidote, and other common heavy minerals in the sediments (Xu and Li, 2003). In recent years, human activities play an important role in the transformation and evolution of landforms in Xiamen Bay. With the construction of seawall, dock and cofferdam, the area of sea surface has decreased greatly. The tidal volume of the West Waters decreased by >110 million m^3, about 33% of the tidal prism since 1938 (Wang et al., 2013). To make matters worse, soil erosion and construction activities during coastal engineering, especially the reclamation without cofferdam, have increased the sediment concentration of seawater, reducing scouring capacity in the bay. Some of the shoals, waterways and anchorages have been seriously silted up.

With a subtropical oceanic monsoon climate, the monthly average air temperature in Xiamen was the lowest (10°C) in January and highest (32.2°C) in July, according to the data from Xiamen meteorological station during 1981–2010 (China Meteorological Data Service Center, 1981–2010). The extreme temperatures were 2.0°C and 39.2°C, respectively. The precipitation recorded in Xiamen was between 67.9 and 552.6 mm. The lowest monthly average precipitation (28 mm) appeared in December, while the highest (207.1 mm) appeared in August. From September to March the northeast monsoon is the most active, while during April and August the southeast monsoon is dominant.

According to the statistical wind data of Xiamen meteorological station during 1980–1999, the easterly wind often dominated throughout the year in Xiamen Bay, with the annual average wind speed of 3.8 m/s. On average, there were 27.7 strong wind days every year, with a wind speed >10.8 m/s. The maximum wind speed was 23 m/s. Xiamen Bay is often affected by typhoons. With the data given in Typhoon Almanac from 1949 to 1988 and Tropical Cyclones Almanac from 1989 to 2000, the number of tropical cyclones (with maximum wind speed >11.9 m/s) within 500 km from Xiamen Bay was 6.7/year on average. The number of tropical cyclones with the maximum wind speed >24.5 m/s was 4.2/year, while that with the maximum wind speed >32.7 m/s was 3.7/year. Frontal rain was common from April to June, while rainstorm was more frequent from July to September. Typhoon rainfall was higher than frontal rainfall from April to June. The most severe typhoon in decades was Typhoon Meranti, which landed in Xiamen on September 15, 2016. The wind speed during its landfall was between 42.0 and 50.0 m/s. The measured maximum instantaneous wind speed reached 63.7 m/s, and the central minimum pressure was about 945 hPa. Wind waves are the main form in the bay. The frequency ratio of wind waves to swells during January 1978 and December 1980 ranged from 89 to 11 (Fang, 2008). Annual sunshine duration in Xiamen is 1826.9 h on average (2016 Yearbook of Xiamen Special Economic Zone, 2016). The water temperature in Xiamen Bay is above 20°C in most of the year except for winter, which is suitable for the growth of plankton. As the spring season arrives, the rapid warming creates excellent conditions for algae outbreak. The water temperature in the estuary is slightly higher than that in the bay mouth, for example, the annual average water temperature in the estuary is 29.4°C in summer (the temperature in summer of 1999 was 28.9°C), and is about 15.4°C in winter (Maskaoui et al., 2002; Ma et al., 2004). The average water temperature around Kinmen Island (near the bay mouth) is 26.3°C in summer, and is only 13.9°C in winter. The observed data also indicate that the closer to the bay, the lower the water temperature, but the changes caused by this geographical differences are ignored compared to the effect of weather and solar radiation.

With the geographical barriers such as Kinmen Island, and islets like Da'dan, Er'dan, Qingyu, and Wuyu, the hydrodynamics in the bay are complex. The Taiwan Strait nearby is affected by highly seasonal change of the Zhe'min Coastal Current, a branch of the Kuroshio Current and the South China Sea Warm Current. Due to the semienclosed topography, the seasonal change of marine environment in the bay is relatively weak. Influenced by the Jiulong River, the salinities in different depths (0–12 m) near the bay head change obviously, varying from 3 to 24 during the low tide to 9–30 during the high tide (Wang and Jiang, 2013). The minimum salinity could be below 1 in summer (Maskaoui et al., 2002) and 12 in winter (Wang et al., 2015). The salinity near Kinmen Island has little seasonal change (29 in summer to 31 in winter 2014), with most being higher than 26, except for the one obtained during summer neap tide (23.74).

2 Biodiversity and Ecological Disasters

The history of biodiversity research in Xiamen Bay is long, resulting in abundant survey data. Influenced by the fresh water from the rivers, the seasonal variation of the species proportion is obvious. Since the first report published in 1860, 5713 species of marine/coastal organisms in Xiamen Bay have been recorded, which belong to 41 phylum, including 1390 plankton, 1960 benthos and 670 nekton (Huang, 2006). Animalia accounts for 64.22% of all species; and Protista, 23.67%. Arthropod (928 species), Chordate (909), diatoms (684), and mollusks (558) are the top four (Huang, 2006). Some species in Xiamen Bay were initially discovered and named after the locations or people, such as Jiageng Jellyfish (*Acromitus tankahkeei* Light) and Xiamen Sinoflustra (*Sinoflustra amoyensis*). Since 1956, significant changes have taken place in the marine ecosystem of Xiamen Bay, for the waters such as those in Xinglin Bay, Maluan Bay, and Yundang Lake became enclosed or semienclosed. From 1954 to 1980, the number of certain *Corethron* decreased by half. In highly polluted Maluan Bay, both species and their quantity are fewer. Dolphin, lancelet, limulus, and egret became rare and endangered marine species in the last four decades. Therefore, the government has set up national protected areas for these fauna, and supported conservation research. There are 23 aquatic species in Xiamen Bay, which are protected by the Wild Animal Conservation Law of the People's Republic of China. Among them, Chinese sturgeon (*Acipenser sinensis*) and Indo-Pacific humpback dolphin (*Sousa chinensis*) are Class I protected species; so are other 10 species including finless porpoise (*Neophocaena phocaenoides*). Eleven other species are Class II protected species. In addition, there are three species including Threespot seahorse (*H. trimaculatus*), which are protected by the Convention on International Trade in Endangered Species of Wild Fauna and Flora, Appendices I, II and III, CITES, 2000/034, and viewed as Class II protected species (Huang, 2006).

The species in Xiamen Bay can be divided into three groups geographically, namely, the coastal eurytopic species (the species adapted to high temperature and high salinity) that dominates in the Kinmen Waters, the native species occupying the bay and the freshwater-dominant species. Oysters, shellfish, algae, and vascular plants are distributed on beaches, and the dominant species in high, middle, and low-tide areas show obvious differences owing to the huge tidal range. The sedimentary environment and the nutrient input from upstream bring about the natural mangrove wetland ecosystem in the Jiulong Estuary and Xiamen Bay coast. The artificial restoration of mangrove is also fruitful. For now, eight species including *Kandelia candel, Aegiceras corniculatum* (L) Blanco and *Avicennia marina* (Forsk.) Vierh have been found in Haicang Bay, Xinglin Bay, Tong'an Bay, Wuyuan Bay, and several islets. Historically, Xiamen Island was named after *Ardeidae Egretta* (Egret Island). In 1860, a heron (*E. eulophotes*) was first found in Xiamen and published by an Englishman named Swinhoe. The number of egrets had been reduced with decreased wetland areas, but has rebounded in recent years. By now, 10 species of *Ardeidae* with the population over 30,000 have been found in Xiamen Bay, accounting for 50% of the total species in China. All five species of *Egretta* found in China can be found in Xiamen. Besides, the wetlands in Xiamen Bay are overwintering grounds of winter migrants like cormorant (*Phalacrocorax carbo sinensis* (Biumeenbach)). In winter, simultaneous fishing process of thousands of cormorants can be observed. A total of 313 species of *Aves* in Xiamen Bay have been recorded, many of them being protected by the Wild Animal Conservation Law (Huang, 2006).

Among China's tourist cities, Xiamen is known for its marine ecology diversity. The most famous species are dolphins, seabirds and mangroves. Small cetaceans such as *Delphinidae* and *Phocoenidae* are often found in the bay, but only a few large whales could swim from the Taiwan Strait into the bay. The Indo-Pacific humpback dolphin is one of the most common aquatic mammals (Thomas and Samuel, 2004). As emblematic of smooth sailing in Xiamen's custom, it is the most welcomed local species. Most of them were found in the bay or in the estuary, especially near the West Waters and Tong'an Bay. The number of Indo-Pacific humpback dolphin had decreased substantially due to the enabled waterways and settled nets. But the establishment of protected areas and the reduction of aquacultures have changed this decline trend. A survey shows that from February to December in 2004, the population of Indo-Pacific humpback dolphin in Xiamen Bay reached about 60–80. These dolphins have become an important landscape again. On the other hand, Xiamen Bay is the major production area of lancelet (*Branchiostoma belcheri*), especially in Tong'an Bay and the East Waters. Lancelet was over-fished, so commercial fishing has been banned since 1969. The population is rising through artificial nursing and stock enhancement. Before the 1980s, *Tachypleus tridentatus* distributed in Tong'an Bay, but it is hard to be found now (Chen, 2016). Stock enhancement of limulus has been widely tried in Xiamen, but the effect is not significant. For now, there are 46 main marine breeding species in Xiamen Bay, including 22 fishes such as snapper, nine crustaceans such as mangrove crab (*Scylla serrata*), 12 mollusks such as *Sinonovacula constricta* (Lamarck), and three macroalgae such as *Porphyra haitanensis*. Among them, the majority production is shellfish. In recent years, the aquaculture scale of Xiamen has been gradually reduced. Based on the survey of fishing in Xiamen Bay, fish, shrimp, crab, squilla, and cephalopods are common, and the dominant species are common species such as *Johnius grypotus*, *Charybdis variegate* and *Charybdis japonica*. In spring 2012, the numbers of eggs and larvae in the West Waters were 47.78 ind/100 m^3 and 42.76 ind/100 m^3, respectively; in autumn 2012, the numbers were 1.50 ind/100 m^3 and 4.08 ind/100 m^3, respectively. In spring 2014, the numbers in the waters near Tong'an Bay and Wuyuan Bay were 25.4 ind/100 m^3 and 37.2 ind/100 m^3, respectively; and in autumn 2014, they were 9.40 ind/100 m^3 and 0.40 ind/100 m^3, respectively.

Affected by human activities, there are a lot of exotic invasive species in Xiamen Bay, among which 26 aquatic species and >38 coastal species have been identified (Huang, 2006). The most prominent one was *Spartina alterniflora*. It was introduced in the 1980s to protect wetlands from erosion, and had multiplied rapidly due to the destruction of mangrove areas at the turn of the century. With the government-sponsored weeding activities, this species has been competitively inhibited by local species like *Kandelia candel*. The second most important exotic species was *Mytilopsis sallei*, which had crowded out local shellfish, and soon farmers found it fit for mud crab feeding (Huang, 2006). Its number is controlled effectively now, though it is still widely distributed.

The nutrient flux from the Jiulong River has increased gradually in recent years. The average dissolved inorganic nitrogen flux of the Jiulong River in 2009 was only 32.4 mol/s, dissolved active phosphorus flux was 0.3 mol/s and dissolved silica flux was 45 mol/s. These values were far lower than those during 2010 and 2011 (108.0, 0.7 and 117.1 mol/s, respectively). In view of the fact that the N/P ratio is much higher than the Redfield value in Xiamen Bay, the increase of phosphorus load may cause the eutrophication trend in the bay. It leads to high chlorophyll *a* and primary productivity in Xiamen Bay. For instance, in spring 2012 the mean chlorophyll *a* content in the West Waters was 2.65 mg/m^3, the mean primary productivity was 128.0 mg C/(m^2·d) and the phytoplankton density was 1.53×10^4 cells/L; in autumn 2012, these values were 3.66 mg/m^3, 364.8 mg C/(m^2·d) and 2.22×10^5 cells/L, respectively. Meanwhile, in the waters near Tong'an Bay and Wuyuan Bay in spring 2014, the mean chlorophyll *a* was 2.10 mg/m^3, the mean primary productivity was 168.7 mg C/(m^2·d) and the phytoplankton density was 3.05×10^4 cells/L, while in autumn 2014, these values were 1.83 mg/m^3, 140.8 mg C/(m^2·d) and 3.57×10^4 cells/L, respectively. Among them, diatom was dominant in both species composition and abundance. Eutrophication resulted in the change of phytoplankton community structure and the frequent occurrence of red tide in Xiamen Bay, especially during spring and summer. Compared with the historical record, small phytoplankton (nano-phytoplankton) takes advantage in Xiamen Bay in autumn. Smaller pico-phytoplankton began to dominate in the West Waters, where the most severe eutrophication level was present (Fu et al., 2016). Lower chlorophyll *a* and phytoplankton density ratio in the West Waters in autumn confirmed this. At present, there are at least 105 species of red tide organisms found in Xiamen Bay, most of which are unicellular protists belonging to six categories such as dinoflagellate and diatom. Among them, twelve species are toxic. The red tide species dominated in Xiamen Bay these years is *Akashiwo sanguinea*, different from the species in the historical reports such as *Skeletonema costatum*. It implies that the changed environmental conditions and exotic species intrusion have caused the change of plankton community structure in Xiamen Bay. In recent years, diarrhetic shellfish poisoning has been detected in mollusks occasionally, though none reached the toxic level. To warn citizens of red tide disaster in a timely manner, the managers in Xiamen City are building a red tide warning platform.

3 Human Interventions: Reclamation and Its Environmental Impacts

3.1 Extensive Reclamation Works in Xiamen Bay From the Year 1955

In past decades, socioeconomic development was confined by land shortages. Then, coastal reclamation became a popular way to solve the supply-and-demand imbalance of land resources in China. Fig. 2 provides a clear illustration of the coastline changes between 1955 and 2013 due to extensive reclamation works that have been carried out in Xiamen Bay.

Xiamen Island was originally surrounded by seawater until 1955 when the government launched the construction of Gaoji Dike. By building Gaoji Dike, the Xiamen government aimed at improving the railway and highway transportation of Xiamen City. The dike connects Xiamen Island to the mainland and hence improves public transportation. However, it divides Xiamen West Harbor and Tongan Bay into two isolated zones, leaving only a narrow channel of ~13 m wide. Another seawall, Xinglin Dike, was built in 1956. However, it totally separates Xinglin Bay from Xiamen West Harbor.

To develop the salt industry, the Xiamen government built a dam in the mouth of Maluan Bay in 1960. The dam then separated Maluan Bay from Xiamen West Harbor. Then in the 1970s, Yundang Dike, known as the seawall from Fuyu Islet to Dongdu, was completed. As a result, Yundang Port has been changed into an inland lake, namely, Yundang Lake. After that, Dongkeng Dike, a dam standing in the mouth of Dongkeng Bay, was constructed.

After 1984, several reclamation works were successively carried out due to the construction projects at Xiamen Airport, Dongyu Bay, Wuyuan Bay, Zhangzhou Harbor, Haicang Port, and in Wutong and Dadeng areas (Zhang, 2008).

By comparing the sea charts of different years, Lu (2010) calculated the area of reclamation in Xiamen Bay in each period. Up to the year 2010, the total area of reclamation in Xiamen Bay had reached ~140 km². Xiamen West Harbor and Tongan Bay became

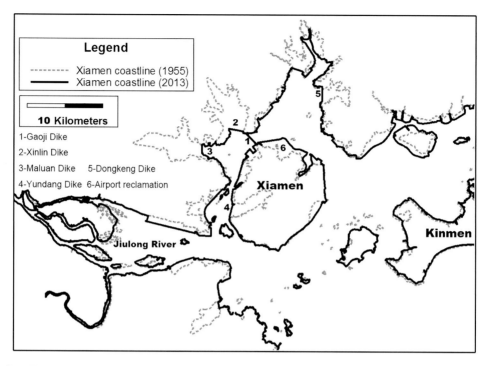

Fig. 2 Coastline change due to land reclamation in Xiamen Bay between 1955 and 2013.

two main regions of reclamation, accounting for 53% and 30% of the total reclamation area, respectively. For several decades, land reclamation had been an effective way to alleviate land shortages. Today, the area of reclamation is still growing in China (Zhu and Xu, 2011). However, according to the existing studies, land reclamation has exerted negative effects on marine ecosystems in many aspects, as we will discuss next.

3.2 Environmental Impacts

Scientists started research on reclamation and its environmental impact on Xiamen Bay as early as the 1980s, and these studies were mainly based on observational data in early years. From August 1980 to September 1981, the Third Institute of Oceanography (TIO) under the State Oceanic Administration of China (SOA) conducted an integrated investigation of marine environment of Xiamen Port. Based on the survey data and sea charts of different periods, Liu et al. (1984) reported that Xiamen Bay is a typical tidal inlet system and that tidal current is the dominant force for shaping its deep navigation channel. He suggested that coastal reclamation resulted in a decrease of tidal prism and hence the drop of current velocity, which further resulted in a reduction of sediment-carrying capability and siltation in the deep channel. Thus, to reduce siltation and keep water depth relatively stable, reclamation works should be carefully investigated before launched, and those that go against deep channel maintenance should be prohibited. Du (1992) summarized the reclamation works during the previous 35 years. He suggested at least 12 million m^3 of the tidal prism in Xiamen Bay was reduced. Consequently, the tidal current velocity decreased and the sediment deposition rate increased. Yang (1994) investigated the submarine scouring-silting variance by comparing submarine topographic profiles in different periods and analyzing the distribution of grain size of sediments. He drew a conclusion that, due to the weakened tidal dynamics, the siltation rate of the Songgu Channel accelerated during the previous decade. In 2004, Lin et al. (2004) showed that most natural shores of Tong'an Bay have been replaced by artificial coast and that some tidal flats are undergoing slow silting evolution based on remote sensing data.

In 2005, an integrated investigation and assessment project of coastal marine environment of Fujian Province (the "Fujian 908 Project") was launched. A great amount of hydrometeorological, chemical, and biological data was obtained and analyzed. Bao et al. (2008) indicated that extensive reclamation in Xiamen Bay during the previous half century had led to the great decrease of environment capacity and to the degradation of water quality and sedimentary environment. The marine ecosystem has been changed due to intense human interventions. As a result, the diversity of species has declined. Following evidence also confirms this trend. The Bingzhou Waters, located at the top of Tongan Bay, are well known as the habitat of the Chinese white dolphin (scientific name: *Sousa chinensis*). However, due to the extensive reclamation in this area and enhanced land-based pollution, the main active region of these precious and rare dolphins has shifted to other water bodies (Liu and Huang, 2000). Around the 1950s, the Liuwudian Waters were famous for abundance of Branchiostoma; however, the number of Branchiostoma decreased rapidly after the 1970s. Lin (2006) suggested that land reclamation had changed the sedimentary environment of Tongan Bay from sandiness to muddiness. This change led to destruction of the Branchiostoma habitat. There is also a remarkable shrinking of mangrove area due to land reclamation (Lin et al., 2006). In the 1960s, the total mangrove area of Xiamen Bay was about 320.0 km^2, but it dropped to 106.7 km^2 in 1979, to 32.6 km^2 in 2000 and to only 21.0 km^2 in 2005, at a time when the mangrove around Xiamen Bay was very close to extinction (Lin et al., 2005).

In the 2000s, the principals of environmental economics were initially applied to assess the environmental impacts of land reclamation on Xiamen Bay. Based on the ecological footprint (EF) method, Meng et al. (2007) indicated that the reclamation projects had resulted in ecological capacity reduction of 77,420.54 hm^2. Economic valuation of ecosystem damages is an important building block in the development of environmental full-cost accounting, which will lead to great improvement in environmental policy-making. A monetary valuation model was established to estimate the losses of coastal ecosystem services due to sea reclamation projects. The result showed that from 1980 to 2005, the cumulative losses of marine ecosystem services caused by reclamation in Xiamen Bay were over 780 million RMB (Lin et al., 2010). Wang et al. (2010) developed a framework for the selection of relevant valuation methods for ecosystem services, which was then applied to estimate the total ecosystem loss due to land reclamation. His results indicated that the costs associated with ecosystem damages are significantly higher than the internal costs of the reclamation projects in Tong'an Bay.

In recent years, numerical modeling has become an effective tool to quantify the impacts of land reclamation on Xiamen Bay. Lu (2010) applied the MIKE21 hydrodynamic model to quantify the cumulative effects of costal reclamation. In his results, the tidal prism passing through the Songyu-Gulanyu section, the Xiamen-Gulanyu section and the Xunjiang section decreased by 54.9%, 24.5% and 23.7%, respectively, by 2009. The coastal reclamation also changed the flow velocity in Xiamen Bay remarkably, especially the flow velocity in the main channel and the mouth of Xiamen Western Sea and Tongan Bay. The flow velocity in the Songyu-Gulanyu and Xiamen-Gulanyu channels decreased by >40% and 20%, respectively. Wang et al. (2013) obtained similar results using the Princeton Ocean Model (Blumberg and Mellor, 1987). According to the model results, the hydrodynamic conditions of Xiamen Bay continuously declined from 1938 to 2007. Specifically, both tidal current speed and tidal prism decreased by 40% in the western part of the bay and by 20% in the eastern part. Nonetheless, Wang et al. (2013) suggested that the hydrodynamic conditions can be restored to certain extent with the implementation of a sustainable coastal development plan, though a full reversal seems not possible. To provide scientific support for the implementation of a coastal reclamation restriction mechanism, Peng et al. (2013) established an analytical framework to estimate the Total Allowable Area for Coastal Reclamation

(TAACR). The result showed that the TAACR in Tongan Bay is 5.67 km², and the area of the bay should be maintained at no <87.52 km². In fact, the area of reclamation in Tongan Bay was >30 km² in 2004, which is far more than the modeled TAACR (5.67 km²) (Lin et al., 2004). This, to certain degree, indicates the overdevelopment of the Xiamen coastal zone.

4 Modeling Approach

4.1 Introduction of the Environmental Fluid Dynamics Code (EFDC): A Powerful Tool for Coastal Environmental Management

The Environmental Fluid Dynamics Code (EFDC) (Hamrick, 1992a) consists of a three-dimensional (3D) surface water modeling system for hydrodynamics, sediment transport, and water quality simulations of lakes, reservoirs, rivers, wetlands, estuaries, and coastal oceans. This integrated modeling system was originally developed at the Virginia Institute of Marine Science in the United States, aiming at providing an operational model for environment and resource management in Virginia's estuarine and coastal waters (Hamrick, 1992b). It also offers a toolkit for fundamental physical and biogeochemical research in aquatic systems (Hamrick and Wu, 1997).

The code solves for 3D primitive variables vertically using hydrostatic equations of motion for turbulent flow in a coordinate system. The system is curvilinear and orthogonal in the horizontal plane, and is stretched to fit bottom topography and free surface displacement in the vertical direction, which is aligned with the gravitational vector. A second moment turbulence closure scheme relates turbulent viscosity and diffusivity to turbulence intensity and a turbulence length scale (Mellor and Yamada, 1982). Transport equations for turbulence intensity and length scale, as well as for salinity, temperature, cohesive and noncohesive sediments, dissolved and adsorbed contaminants, and dye tracers, are solved simultaneously. An equation of state relates density to salinity, temperature, pressure, and suspended sediment concentration. A detailed description of the EFDC and the governing equations can be found in Hamrick (1992a).

The EFDC is also capable of simulating the wetting and drying cycle. It includes a near-field mixing zone module, which is fully coupled with a far-field transport of salinity, temperature, sediment, contaminants, heavy metal, and eutrophication variables. In addition, the EFDC contains hydraulic structure representation, vegetative resistance, and Lagrangian particle-tracking modules. The model also accepts radiation stress field from wave refraction-diffraction module; therefore, it can simulate longshore currents and wave-induced sediment transport.

The EFDC has been continuously developed since 1992. Today, the EFDC has become an integrated water environment modeling system in support of environmental-ecological assessment and management, environment related regulations and science-based policy-making (www.epa.gov/exposure-assessment-models/efdc).

4.2 Application of EFDC to Study Suspended Sediment Concentration in Xiamen Bay

4.2.1 Setting Up a Numerical Model for Xiamen Bay

The EFDC was applied to simulate the hydrodynamics and suspended sediment concentration (SSC) of Xiamen Bay (Xie et al., 2016a,b). The mesh of the hydrodynamic model was formed by 38,862 horizontal orthogonal grid cells covering the area of 117.7°E-118.6°E, 24.25°N-24.7°N, with 10 vertical layers using the sigma coordinate. The horizontal resolution varied between ~80 m to ~300 m, with higher resolution in the inner harbor and lower resolution in the outer harbor.

To force the model at the open boundary, the tidal forcing was used in the coastal area between Weitou and Liuhui. The tidal components were calculated from in-situ sea level data. The semi-diurnal components used to force the model were (M2, S2, N2, and K2), and the diurnal components were (K1, O1, Q1, and P1). Three shallow-water components (M4, MS4 and M6) were also used. For the internal boundary, such as upstream river boundary, the Jiulong River was considered as the only dominant inflow. To reflect the average effect of the Jiulong River, the runoff condition of the upstream boundary was set to the annual average according to statistical data, and salinity was set to zero. With tidal current velocity of up to ~2 m/s, the macro-tides in Xiamen Bay (average tidal oscillation was ~4.00 m with the maximum record of 6.42 m) dominated the transport of sediment (Zuo et al., 2011; Zheng et al., 2013). The additional effects from wind and heat fluxes at the free-surface boundary were negligible, allowing the simulation to be forced by tides and river runoff only.

For the initial conditions of the hydrodynamic module, constant values for salinity (34 psu) and temperature (25°C) were used. The simulation started on 00:00:00, 1st November 2011, with a five-second time step, and a 30-day computation duration.

Fig. 3 shows tidal curve comparison between observation (solid line) and model results (dotted line) at T2 (Station Xiamen) and T3 (Station Weitou). Fig. 4 illustrates the velocity fields of the Jiulong Estuary and Xiamen Bay at the peak ebb and flood during spring.

To calculate SSC in the Jiulong Estuary and Xiamen Bay and to further improve the usability of the model for coastal environmental assessment and management, the hydrodynamic model was simplified based on the characteristics of this coastal area. Xiamen Bay, with general water depth of 3–10 m (which is the lowest of normal low water) and strong tidal currents, is a dynamic shallow water of intense turbulence and mixing. Therefore, only a 2-dimentional sediment transport model was considered in the sediment module.

The sediment in the Jiulong Estuary and Xiamen Bay is mainly formed by fine cohesive materials with the clayey silt accounting for 65.22% (Zuo et al., 2011). Sediment in the outer harbor and around Kinmen Island basically consists of silt with grain size varying from 4.80–7.48ϕ (i.e., 7.6–37.8 µm) (Fang et al., 2010). Larger sediment particles, for example, the sands, are also observed in this area; however, they provide little contribution of SSC in Xiamen Bay. Thus, all kinds of sediments are assumed to behave as fine and cohesive sediments in the model.

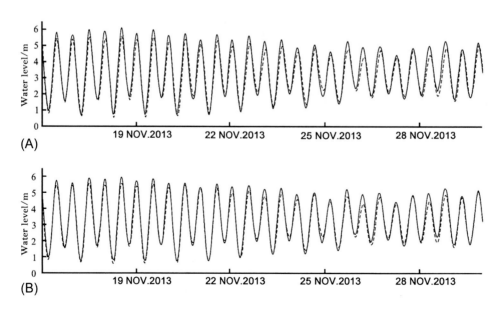

Fig. 3 Tidal curve comparison between observation and modelled T2 (A), and between observation and modelled T3 (B).

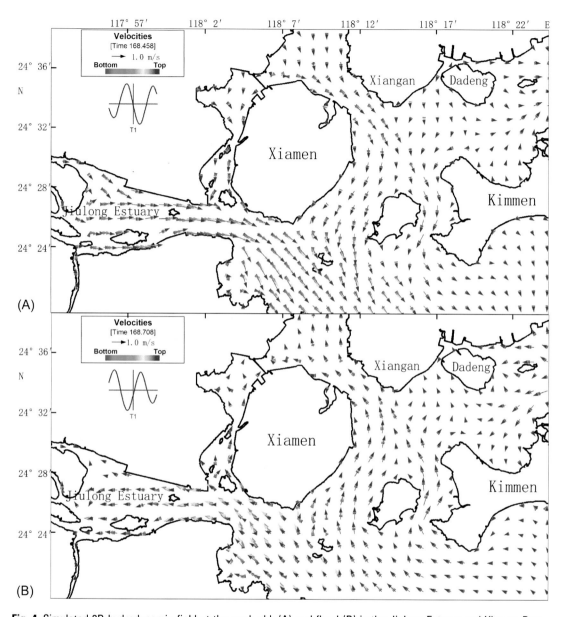

Fig. 4 Simulated 3D hydrodynamic field at the peak ebb (A) and flood (B) in the Jiulong Estuary and Xiamen Bay.

Key parameters of the sediment model are seabed roughness Z_0 (0.01 m), surface erosion rate $\mathrm{d}m_e/\mathrm{d}t$ (0.19 g/(m^{2}s)), critical shear stresses for deposition τ_{cd} (0.0025 m^2/s^2) and for erosion τ_{ce} (0.004 m^2/s^2). The settling velocity W_s was set to be a function of SSC (S). The remaining numerical and physical parameters, initial conditions and equations used in the model can be found in Xie et al. (2016b). To calibrate and validate the sediment model, in-situ data measured in 2013 including sea level, current velocity, and SSC were used. For detailed location information of the observation stations, please refer to Fig. 1.

4.2.2 *Temporal and Spatial Variation of SSC in Tidal Cycles*

Fig. 5 illustrates the temporal variation of SSC during both spring and neap tides at nine observation stations (1#–9#) in and around Xiamen Bay, with a solid line representing the model result and a dotted line, the in-situ data. Fig. 6 gives SSC horizontal distribution at high and low tides during spring and neap tides in the Jiulong Estuary and Xiamen Bay.

The SSC was much higher during spring tide than during neap tide, and highly correlated with tidal periodicity. The SSC amplitude during spring tides was large, especially in shallow waters. For example, the SSC of Station #7 ranged from ~50mg/L to ~250mg/L in a spring tidal cycle and varied even larger from ~80mg/L up to 500mg/L at Station #2. Generally, there were two SSC peaks in one day during spring tide. In conjunction with Fig. 6, it was clear that two SSC peaks were all formed during the ebb, with the SSC maximum around low water. In contrast, the SSC was much lower and more stable during neap tides, and showed weak fluctuation in a tidal cycle. At Stations #1 to #9, SSC remained in a range of ~20–40mg/

As shown in Fig. 6, the SSC gradually increased from the outer bay to the inner bay, with higher concentration in coastal shallow waters. The SSC was the highest in the Jiulong Estuary, and in the Xiang'an and Dadeng waters within the study area, which was consistent with the remotely sensed SSC distribution of Xiamen Bay in Luo et al. (1999) and Lin and Chen (2008).

During spring, a turbidity maximum in the Jiulong Estuary with SSC of ~250mg/L existed to the northwest of Haimen Island at high water. The turbidity maximum then moved seawards during the ebb, and it stayed to the east of Haimen Island at low water with SSC up to 400mg/L. This modeled turbidity agreed well with the observation (Wang, 2005). In addition, a suspended sediment plume with SSC of ~75–150mg/L formed to the east of the Jiulong Estuary, and then extended southeastward, reaching as far as the Liuhui water. The plume reciprocated in and out of the estuary with flood and ebb tides. Note that part of the suspended sediment shifted in the West Xiamen Harbor when the plume retracted with flood tide. This process was also demonstrated in a study using Grain-Size Trend Analysis (GSTA) method (Zuo et al., 2011).

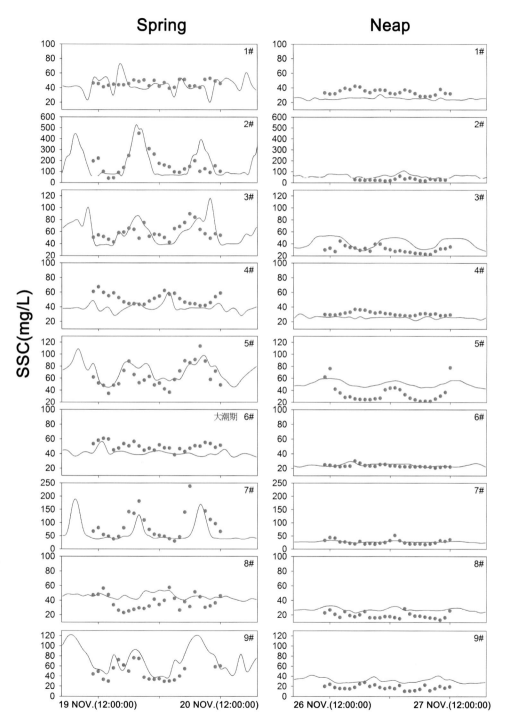

Fig. 5 Comparison of observed *(dotted)* and simulated *(line)* SSC (depth averaged) during spring and neap tides.

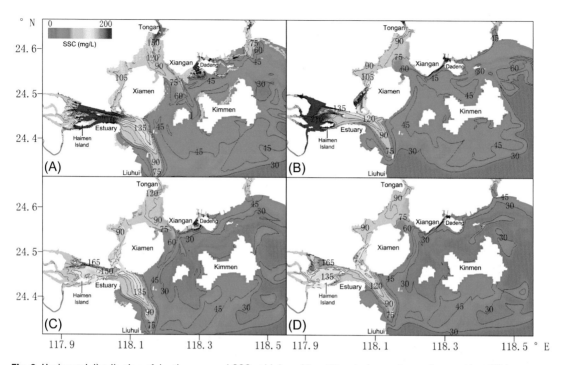

Fig. 6 Horizontal distribution of depth-averaged SSC at high and low tides during spring and neap tides. (A) Low water in spring. (B) High water in spring. (C) Low water in neap. (D) High water in neap.

The simulated SSC in other areas of Xiamen Bay also agreed with previous publications, except for the area to the northwest of Xiamen Island where the SSC is slightly overestimated (Luo et al., 1999; Wang, 2005; Lin and Chen, 2008; Zuo et al., 2011).

4.2.3 Dredging-Produced SSC Increment and Marine Pollution

In and around Xiamen Bay, there are wildlife conservation areas and tourism functional zones (see Fig. 7), for example, the Chinese white dolphin nature reserve in the Xunjiang water, the Xiamen Branchiostoma conservation between Kinmen and Dadeng, the Xiamen rare and endangered species protection zone located to the south of Xiang'an and Dadeng, and a tourism functional zone along the eastern Xiamen coast. In recent years, however, to support the coastal reclamation projects in Xiamen, extensive dredging works have taken place in Xiamen Bay, especially in the shallow water near Xiang'an and Dadeng areas. Given that there are important conservation zones in this area, the dredging-produced SSC increment and

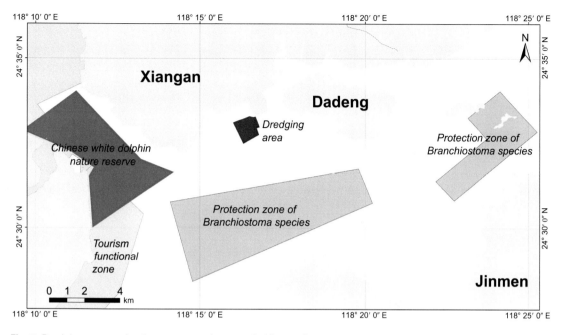

Fig. 7 Dredging area and adjacent protection zone in Xiamen Bay.

the resulting marine pollution have become major issues for Xiamen Bay environmental management.

A dredging project conducted near the eastern Xiang'an coast and the consequent SSC increment were investigated using the Xiamen Bay EFDC (Xie et al., 2016a,b). Fig. 8 shows the long-distance transport of dredging-produced sediment. We can see that during ebb, a part of dredging-produced suspended sediment is "pumped" off the southern Xiang'an coast toward the protection zone of Branchiostoma species. During floods, most of these suspended sediments, together with those near the dredging area, are shifted into the Xunjiang water, which causes an SSC increment around the Chinese white dolphin nature reserve. Note that a part of the SSC in the Xunjiang water can be transported to the eastern Xiamen coast by ebb tidal currents from Tong'an Bay and the Xunjiang water. As a result, a zonal distribution of SSC forms around eastern Xiamen Island, which happens to cover most of the tourism functional zone.

Although the SSC increment is not large under the planned workload, the region influenced by the SSC increment is extensive. In fact, the dredging-produced SSC can be transported far away from the dredging area by strong tidal currents, and then it may cause pollution and damage to marine ecosystems at a distance.

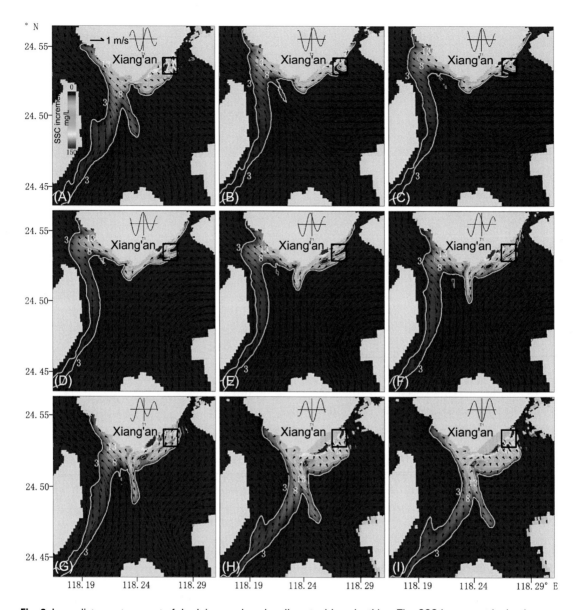

Fig. 8 Long-distance transport of dredging-produced sediments driven by tides. The SSC increment is depth-averaged, HW represents high water, LW represents low water, and h stands for hour. *Black arrows* represent depth-averaged current velocity. A *black square* marks the dredging area, which corresponds to the dredging area marked in Fig. 7. (A) HW-3h. (B) HW-1h. (C) HW. (D) HW+0.5h. (E) HW+1h. (F) HW+2h. (G) LW-2h. (H) LW-1h. (I) LW.

In a case study, when the dredging workload, namely, the SSC source intensity, was tripled in the Xiamen Bay EFDC, water pollution broke out in the protected zones mentioned above (e.g., SSC increment >30 mg/L in the Chinese white dolphin nature reserve).

5 Recommendations and Future Steps

To build a blue bay with "clean seawater, pure sand beach, green coast, beautiful bay, and amazing island," Xiamen government has proposed many pollution control measures and comprehensive remediation projects, which were carried out in recent years. They included eutrophication pollution control marine stock enhancement release, reconstruction of mangrove ecosystem, eco-restoration of uninhabited islands, construction of national marine park, ecological transformation of coastline, reduction of marine debris, and construction of marine reserves. Based on these projects, Xiamen government also put forward policies, plans, and eco-compensation mechanisms to improve the quality of seawater and mitigate human interventions for the blue bay remediation.

(1) Governing the sea through laws

Based on related laws and regulations, the local government needs to further modify, improve, and develop clear and complete operational environmental regulations and technical standards for ruling the sea through laws. Those who use sea areas must go through the formal approval process in accordance with the laws, which can stop the overuse of the water and illegal development activities of marine resources according to national ownership of sea use.

Raising awareness for sustainable development is needed at all levels. To attach great importance to marine environment protection and conservation, cleanups in Xiamen have been organized by the China Mangrove Conservation Network (CMCN) and Volunteers Union since 2013. There are >1000 volunteers involved in the action. A monthly Coastal Cleanup and Monitoring Project has been carried out in 12 cities around the bay, including Xiamen.

The comprehensive law enforcement system should be established to strengthen law enforcement including environmental protection department and related management organizations. Environmental information system including database, monitoring system and GIS should be established for integrated management of marine environment and for decision-making services.

(2) Scientific planning and strategic environmental assessment

To achieve harmonious relationship between marine environmental protection and development, and to ensure that each piece of sea area has a scientific rational use and layout, the scientific coastline development plan, marine function zoning, marine development, and protection should be incorporated into the national economic and social development plan and stage plan as legal forms to be confirmed to safeguard their legal status. An example of such planning is Wuyuan Bay. In the past, Wuyuan Bay was an unsuitable living place with large tidal flats and serious deposition. Now, it is a new landmark of Xiamen with clean water, sailing club and wetland park.

Marine ecosystem is highly complex. Each element of the system is in close relation with the others, so "the butterfly effect" should not be neglected. Through the history of dike construction in Xiamen and the assessment of marine environmental impact, a big picture of human interventions and the consequent responses of the marine ecosystem is shown. First, coastal reclamation results in great changes in coastline and topography. As a result, the tidal prism of Xiamen Bay is reduced. The reduction of tidal prism then causes a decrease of current velocity. The sediment-carrying capability of the tidal current, which is proportional to current velocity, is therefore weakened. In this way, the sedimentary environment is changed, and the change then ruins the habitat of Branchiostoma. Eventually, both biodiversity and ecological services degrade. The chain reaction has become a major issue of environmental management.

To deal with this problem, a top-level design and strategic environmental assessment are needed. Through the top-level design, Xiamen Bay is considered as a whole system. Most elements are taken into account to optimize the development and management of the system. By carrying out strategic environmental assessment, most environmental impact factors in the system and possible chain reactions are assessed to minimize the total damage. Today, as the coastal engineering projects are becoming more and more intensive, they should be regulated under a comprehensive and optimized framework.

(3) Improving environmental capacity

To improve marine hydrodynamic and pollutant diffusion conditions and the environmental capacity of Xiamen waters, removing aquaculture, dredging, transformation projects of seawalls, and other ecological restoration projects have been implemented gradually. In the West Waters of Xiamen Bay, $70 \, km^2$ sea area has been completed for remediation, which has cost US$40 million since 2002 involving eight towns, four districts, and >31,000 people. >55,000 mariculture cages were dismantled; 1840 ha marine aquaculture areas were cleared. In the East Waters, the same measures like those used in the West Waters were implemented.

Before remediation, there were many marine aquaculture areas. Eutrophication problem was serious in Xinlin Bay. Now, the water is clean after remediation. By now, over one million square meters of artificial beach have been built, and 30 km of shoreline has been remediated. And 0.16 billion cubic meters of sediment has been dredged. Aquiculture areas decreased greatly in both West Waters and East Waters.

The comprehensive remediation projects in the next two years will be carried out in Xiatanwei coastal wetland and Haicang Bay. Meanwhile, about 6.4 km of beach was remediated, with the mangrove planting area of 0.25 million square meters. The project in the Xiatanwei Wetland will cover an area of 1.7 million square meters, with dredging waterway over 7.3 km. Mangrove plants were densely planted on two artificial islands, with island cofferdam over 3.5 km. The project in Haicang Bay will dredge in 6.66 km^2, with 21.93 million cubic meters sediment. Two seawalls in Xiamen have been demolished, one has been converted, and the other will be converted in 2017. Over 50 batches of larval shrimp and fish were released. Environment impact assessment (EIA) has been conducted currently to avoid negative effects.

The utilization efficiency of environmental capacity of Xiamen sea area will be improved by putting the discharge of tailwater into deep water through location adjustment of discharge port of the sewage treatment plant and construction of relevant pipe network based on the optimal layout of the discharge port.

(4) Other measures to mitigate human interventions

With high population and input of the Jiulong River, there are serious eutrophication problems in Xiamen waters. The seawater quality of most waters in Xiamen is worse than Class IV standard. The use of cleaner production technology should be encouraged to reduce the amount of industrial sewage and to reuse it. Domestic sewage should be treated to reduce nitrogen and phosphorus emissions efficiently by municipal sewage systems according to the level of reuse. Various types of emission reduction targets should be designed scientifically and enforced strictly to reduce pollutants' discharge from the Jiulong River, the Dongxi River, the Xixi River, and other river basins.

It is important to establish an eco-compensation mechanism based on the coordination between downstream and upstream governments to protect the ecological environment of the Jiulong River from the source to the estuary.

To establish marine debris service platform including methods, policy, research, monitoring, assessment, and international cooperation in Xiamen will help to monitor the sources, reduce the amount, and eventually prevent the emergence of marine debris.

References

Bao, J., 2011. Study on Topography and Geomorphology of Xiamen Bay and its Adjacent Regions. Third Institute of Oceanography, p. 55. Master Dissertation, Feng Cai (Supervisor). (in Chinese).

Bao, X., Qaioe, L., Yu, H., 2008. Hydrodynamic Impact Assessment of Reclamation Planning in the Bays of Fujian Province. Science Press, Beijing.

Blumberg, A.F., Mellor, G.L., 1987. A description of a three-dimensional coastal ocean circulation model, three-dimensional coastal ocean models. Coastal Estuarine Sci. 4, 1–16. https://doi.org/10.1029/CO004p0001.

Cai, A., et al., 1993. Research on the sedimentary and acoustic characteristics of new deposit sequence in Jiulongjiang estuary and Xiamen Bay. J. Xiamen Univ. 32 (3), 345–350 (in Chinese).

Chen, J., 2016. Beijing. In: Comprehensive Investigation and Assessment in Offshore of Fujian Province, first ed. Science Press, p. 523 (in Chinese).

China Meteorological Data Service Center, 1981–2010. http://data.cma.cn/data.

Du, Q., 1992. Land reclamation and its impacts on navigation in western Xiamen harbor and fishery. J. Fujian Fish. 02, 85–92.

Fang, J., 2008. Distribution Characteristics, Sources, and Sedimentary Environmental Indications of the Sediment in Xiamen Bay. Third Institute of Oceanography, p. 88. Master Dissertation, Jian Chen (Supervisor). (in Chinese).

Fang, J., Chen, J., Li, Y., et al., 2010. Study of modern sedimentary environment in the Xiamen Bay. Acta Sedimentol. Sin. 28 (2), 356–364.

Fu, T., et al., 2016. Size structure of phytoplankton community and its response to environmental factors in Xiamen Bay, China. Environ. Earth Sci. 75 (9), 1–12.

Hamrick, J.M., 1992a. A three-dimensional environmental fluid dynamics computer code: theoretical and computational aspects. In: Special Report No. 317 in Applied Marine Science and Ocean Engineering. vol. 63. College of William and Mary, Virginia Institute of Marine Science, Gloucester Point, VA.

Hamrick, J.M., 1992b. In: Estuarine environmental impact assessment using a three-dimensional circulation and transport model. Proceedings of the 2nd International Conference, American Society of Civil Engineers, New York, pp. 292–303. Estuarine and Coastal Modeling.

Hamrick, J.M., Wu, T.S., 1997. Computational design and optimization of the EFDC/HEM3D surface water hydrodynamic and eutrophication models. In: Next Generation Environmental Models and Computational Methods. Society for Industrial and Applied Mathematics, Philadelphia, PA.

Hong, H., Chen, F., 2003. Characteristics of heavy minerals in a core from Jiulongjiang estuary. J. Oceanogr. Taiwan Strait 22 (1), 65–78 (in Chinese).

Huang, Z., 2006. Diversity of Species in Xiamen Bay, China, first ed. Ocean Press, Beijing, p. 587 (in Chinese).

Hydrological Information Network of Fujian Province, n.d. http://www.fjsw.gov.cn (in Chinese).

Li, W., et al., 1999. Primary productivity and its relationship with environmental factors in coastal waters off Haicang, Xiamen. Trop. Oceanol. 28 (3), 51–57 (in Chinese).

Lin, G., Fang, J., Chen, F., 2004. A remote sensing analysis of shallow sea topographic evolutionary trend of Tong'an Bay, in Xiamen. Remote Sens. Land Resources 62 (4), 63–67 (in Chinese).

Lin, P., Zhang, Y., Yang, Z., 2005. Protection and restoration of nangroves along the coast of Xiamen. J. Xiamen Univ. Nat. Sci. 44, 1–6.

Lin, Q., Chen, Y., 2008. Multi-temporal analyses of remote sensing on distribution of suspended sediment in Xiamen Estuary. Port Waterway Eng. 12, 51–57.

Lin, T., Xue, X., Shawns, S., et al., 2006. Analysis of coastal wetland changes in Xiamen. China Popul. Resour. Environ. 16 (4), 73–77.

Lin, X., 2006. Xiamen branchiostoma: resource management and economic development. Xiamen Sci. Technol. 1, 13–16.

Lin, X., Chen, W., Rao, H., 2010. Valuation of marine ecosystem service losses caused by sea reclamation: a case study of Xiamen Bay. Ecol. Econ. 2, 385–389.

Liu, W., Huang, Z., 2000. Distribution and abundance of Chinese white dophins (*Sousa chinensis*) in Xiamen. Acta Oceanol. Sin. 22 (6), 95–101.

Liu, W., Tang, Z., Liu, Q., 1984. Submarine geomorphology of the Xiamen harbour and its scouring-silting change. J. Oceanogr. Taiwan Strait 3 (2), 178–188 (in Chinese).

Lu, R., 2010. A Study on the Cumulative Effects of Coastal Reclamation to Hydrodynamics in Xiamen Bay. Third Institute of Oceanography, State Oceanic Administration, Xiamen, China.

Luo, J., Gong, J., Zhang, X., 1999. Multitemporal analyses of remote sensing on distribution, transportation and sedimentation of suspended sediment in the Jiulongjiang mouth and Xiamen estuary. J. Hydrosci. Eng. 4, 368–379.

Ma, Q., et al., 2004. Spatial distribution of seawater dimethylsulfide in winter of Jiulong Jiang Estuary. Environ. Sci. 25 (5), 47–51 (in Chinese).

Maskaoui, K., et al., 2002. Contamination by polycyclic aromatic hydrocarbons in the Jiulong River estuary and western Xiamen Sea, China. Environ. Pollut. 118, 109–122.

Mellor, G.L., Yamada, T., 1982. Development of a turbulence closure model for geophysical fluid problems. Rev. Geophys. 20 (4), 851–875.

Meng, H., Chen, W., Zhao, S., et al., 2007. Application of the ecological footprint analysis in ecological impact assessment of sea reclamation: Xiamen Western Bay case study. J. Xiamen Univ. 46 (A01), 203–208.

Peng, B., Lin, C., Jin, D., et al., 2013. Modeling the total allowable area for coastal reclamation: a case study of Xiamen, China. Ocean Coast. Manag. 76, 38–44.

Thomas, A.J., Samuel, K.H., 2004. A review of the status of the indo-Pacific humpback dolphin (*Sousa chinensis*) in Chinese waters. Aquat. Mamm. 30 (1), 149–158.

TOP 100 Container Ports in 2016, 2016. Lloyd's List. http://promo.lloydslist.com/landingpages/containerports.

Wang, G., et al., 2015. Net subterranean estuarine export fluxes of dissolved inorganic C. N. P. Si. and total alkalinity into the Jiulong River estuary, China. Geochim. Cosmochim. Acta 149, 103–114.

Wang, J., Hong, H., Zhou, L., et al., 2013. Numerical modeling of hydrodynamic changes due to coastal reclamation projects in Xiamen Bay, China. Chinese J. Oceanol. Limnol. 31 (2), 334–344.

Wang, J., Jiang, Y., 2013. The distribution of salinity and the dynamic process of salt flux in Jiulong River estuary. J. Xiamen Univ. Nat. Sci. 52 (6), 835–841 (in Chinese).

Wang, X., Chen, W., Zhang, L., et al., 2010. Estimating the ecosystem service losses from proposed land reclamation projects: a case study in Xiamen. Ecol. Econ. 69 (12), 2549–2556.

Wang, Y., 2005. Initiative research on the transportation of suspended sediments in Jiulong River Estuary, Fujian Province, at both flood and low water seasons. Ocean University of China, Qingdao, China.

Xie, S., Wang, C., Wang, J., et al., 2016a. Numerical simulation study on 3D tidal flow and salinity in the Jiulong estuary-Xiamen Bay based on the EFDC. Chinese J. Hydrodyn. 31 (1), 63–75.

Xie, S., Wang, J., Wang, C., et al., 2016b. Numerical study on suspended sediment concentration in Jiulong estuary-Xiamen Bay and sediment transport mechanism in tidal inlets with multi-fork. Chinese J. Hydrodyn. 31 (2), 188–201.

Xu, M., Li, C., 2003. Characteristics of heavy minerals composition and distribution in sediment from Jiulong River estuary. Mar. Sci. Bull. 22 (4), 32–39 (in Chinese).

Yang, S., 1994. Submarine scouring-silting variation near Xiamen song islet. J. Oceanogr. Taiwan Strait 15 (1), 94–102.

Yearbook of Xiamen Special Economic Zone, 2016. http://www.stats-xm.gov.cn/2016. (in Chinese).

Zhang, L., 2008. Retrospective Assessment of Environmental Impact of Reclamation Planning in the Bays of Fujian Province. Science Press, Beijing.

Zheng, X., Pan, W., Zhang, G., et al., 2013. Distribution characters of suspended sediment and dynamic analysis in the Xiamen Bay. J. Xiamen Univ. Nat. Sci. 52 (4), 539–544.

Zhu, G., Xu, X., 2011. Research review on environmental effects of the land reclamation from sea. Ecol. Environ. Sci. 20 (4), 761–766.

Zuo, S., et al., 2011. Grain size characteristics of surface sediments and dynamic response in sea area of Jiulongjiang estuary and Xiamen bay. J. Hydrosci. Eng. 4, 74–79 (in Chinese).

Further Reading

Luo, Z., et al., 2008. The three dimensional tidal current numerical model in Jiulongjiang estuary and Xiamen Bay. J. Xiamen Univ. Nat. Sci., 864–868.

7

COASTAL DYNAMICS AND SEDIMENT RESUSPENSION IN LAIZHOU BAY

Zai-Jin You*, Chao Chen[†]

*Ports and Coastal Research Centre, Ludong University, Yantai, China
[†]School of Engineering and Technology, Jimei University, Xiamen, China

CHAPTER OUTLINE
1 **Laizhou Bay   123**
2 **Meteorological Climate   125**
3 **Coastal Dynamics   127**
   3.1 Astronomic Tides and Currents   127
   3.2 Coastal Waves   130
4 **Sediment Transport   132**
   4.1 Sediment Type   132
   4.2 Sediment Setting Velocity   133
   4.3 Sediment Resuspension   134
5 **Human Impact   137**
   5.1 Coastal Land Reclamation   137
   5.2 River Reservoir Construction   138
References   140
Further Reading   141

1   Laizhou Bay

Laizhou Bay is in the south of the Bohai Sea and stretches from the modern Yellow River Delta (37.8°N, 119.1°E) in the west to Qimu Island (37.67°N, 120.65°E) in the east (see Fig. 1). The Bay has a bowl-shaped area with an average water depth <10 m and a maximum water depth of about 18 m. The surface water area of the Bay is about 7000 km², and the coastline is around 300 km in length.

Sediment Dynamics of Chinese Muddy Coasts and Estuaries. https://doi.org/10.1016/B978-0-12-811977-8.00007-8

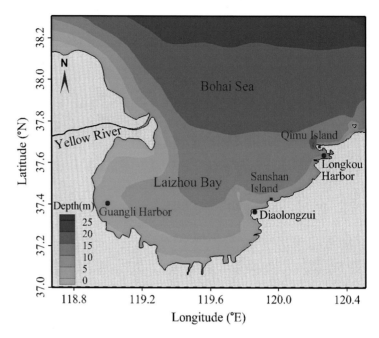

Fig. 1 Location of Laizhou Bay in the Bohai Sea, China.

With the development of coastal economy and maritime industries during the past decades, several harbors were constructed or redeveloped inside the bay, and large-scale coastal land reclamation was carried out along its coastline. Laizhou Bay is also rich in primary fishery production, and it is an important spawning and feeding ground for the fisheries in the Bohai Sea. But Laizhou Bay is part of the Bohai Sea Economic Zone, and it has experienced overfishing, land reclamation, and coastal pollution, resulting in great threats to the bay's biodiversity and ecosystem health.

The Yellow River, one of the most sediment-laden rivers in the world, affects sediment transport and water quality in Laizhou Bay. The river has been discharging into the Bohai Sea on the northwest coast of Laizhou Bay since 1855, and it fundamentally affects the coastal areas adjacent to the river mouth. Tremendous amounts of sediment are transported into the Bohai Sea every year (Wang and Wang, 2010), resulting in the formation and evolution of the modern Yellow River Delta. Since 1960, several large river reservoirs and dams have been constructed in the middle of the Yellow River, leading to dramatic decrease in sediment flux downstream of the river. Furthermore, in scouring the river bed of the lower reaches and alleviating pool infilling in Sanmenxia

Reservoir and Xiaolangdi Reservoir (Yu et al., 2013), the Water-Sediment Regulation Scheme (WSRS) has been in operation by the Yellow River Conservancy Committee (YRCC) since 2002. As a result, the freshwater discharge and sediment load delivered by the Yellow River have been significantly reduced once again. Now, the Yellow River is a highly human-regulated system that affects Laizhou Bay.

Therefore, it is essential to understand hydrodynamics and sediment transport patterns in Laizhou Bay, which are controlled not only by natural physical processes but also by human activities related to protection and management of natural resources in this region. This chapter is organized as follows: this section provides a brief introduction of Laizhou Bay. Section 2 presents meteorological information about the bay. Astronomic tides and coastal waves based on observations and modeling results are included in Section 3. Sediment resuspension dynamics in Laizhou Bay are presented in Section 4. Human impact on Laizhou Bay is finally discussed in Section 5.

2 Meteorological Climate

The overall rainfall in Laizhou Bay is concentrated in summer seasons, and the annual average rainfall is about 800 mm. The hottest month of the year is August, and the coldest is January, and the annual average air temperature is about 12.5°C in the bay.

Fig. 2 shows variations of monthly rainfall and air temperature observed from 1954 to 1980 at two weather stations of Yangjiaogou and Yexian. The overall rainfall is shown to concentrate in the summer season. August is the hottest month, and the coldest month is January. There is no clear difference between the datasets from the two weather stations.

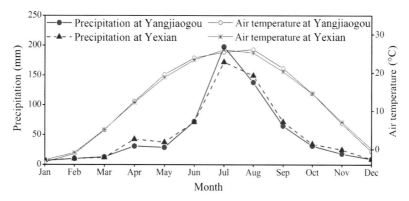

Fig. 2 Observed monthly rainfall and air temperature at two weather stations.

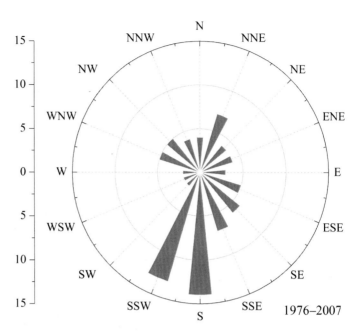

Fig. 3 Statistical frequency distribution (%) of wind directions measured at Laizhou Weather Station over the period of 1976–2007.

Laizhou Bay is generally subjected to the East Asian Monsoon. In winter, strong northerly dominant winds accompany storms induced by cold-air outbreaks, while in summer, weak southerly dominant winds are associated with the Western North Pacific Subtropical High. The frequency distribution of wind directions observed at Laizhou Weather Station over the period of 1976–2007 is shown in Fig. 3. The prevailing wind directions are from the south and southwest, but winds from NNE direction also account for a substantial proportion.

Tropical cyclones and cold-air outbreaks are main drivers for major coastal hazards in Laizhou Bay. In general, damage caused by tropical cyclones is much more serious than that caused by cold-air outbreaks. Fortunately, few tropical cyclones can reach such a high latitude region as Laizhou Bay. Storms related to cold-air outbreaks occasionally attack the Bohai Sea, especially during the seasonal transitions of autumn-winter and winter-spring. Storm surges at high tides cause pronounced damage along the coast of Laizhou Bay. For example, a cold-air outbreak in Laizhou Bay which occurred from March 3rd through the 6th of 2007 resulted in more than US$10 million in fishery and agriculture losses. The cold-air outbreak, which lasted for three days (in Fig. 4), generated strong northerly winds accompanied by

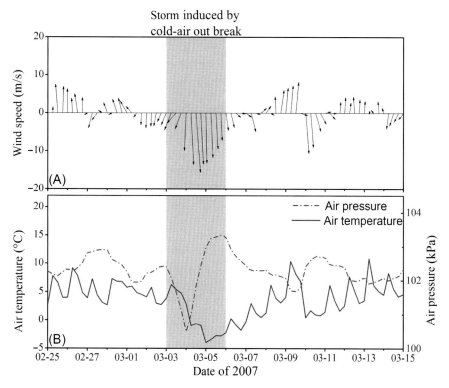

Fig. 4 Cold-air outbreak generated the storm associated with wind speed, air temperature, and pressure data from the ECMWF.

pronounced decreases in air temperature and air pressure, where the data selected in the middle of Laizhou Bay) are from the ECMWF (http://apps.ecmwf.int/datasets/).

3 Coastal Dynamics

3.1 Astronomic Tides and Currents

Astronomic tides in Laizhou Bay are irregular semidiurnal. The average tidal range is 0.92–1.43 m, and it increases from the north to south mainly due to the topography of the bay. Except for the regions close to Longkou Harbor, the duration of ebb tide is longer than that of flood tide. Tidal currents flow southwards during flood tides and move northwards during ebb tides. There are strong tidal currents off the modern Yellow River Delta, and the maximum current speed can reach 1.4 m/s and 1.9 m/s during flood and ebb tides, respectively.

Tidal currents and elevations at different phases in Laizhou Bay are numerically modeled with a three-dimensional, *Finite-Volume Coastal Ocean Model* (FVCOM) (Chen et al., 2003, 2007, 2011). The FVCOM solves the Reynold-averaged Navier-Stokes equations combined with Boussinesq approximation. Modified Mellor-Yamada level 2.5 turbulent closure scheme is used for the vertical eddy viscosity (Mellor and Yamada, 1982; Galperin et al. 1988).

The computational domain in Fig. 5 covers part of North Yellow Sea and the entire Bohai Sea and unstructured triangular grids are utilized in the horizontal to flexibly fit the complex shoreline and islands. A grid size of approximately 1 km is used near the open boundary in the North Yellow Sea, and a finer grid of about 100 m in length is applied in Laizhou Bay. In the vertical, 15 uniform sigma layers are used. The tidal elevation at the open boundary consists of 8 major constituents (M2, S2, N2, K2, K1, O1, P1 and Q1) extracted from the OSU Tidal Inversion Software (Egbert and Erofeeva, 2002).

Tidal currents and elevations at different phases are computed and shown in Fig. 6. The time series of surface elevation at Station 1, which is in the middle of Laizhou Bay and indicated by the solid square, is illustrated on the upper right corner of each figure, and the solid circle represents the present phase in the tide cycle. In Fig. 6A, currents flow southward during the flood tide, and the current speed decreases from the west to east.

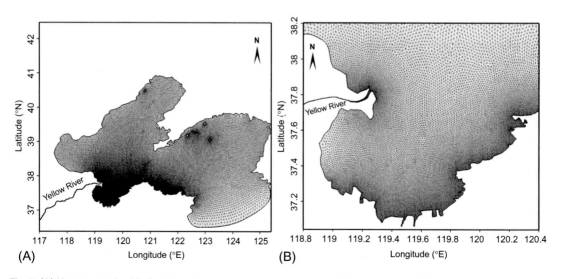

Fig. 5 (A) Unstructured grids for the entire computational domain, and (B) finer grids of 100 m for the computational domain of Laizhou Bay.

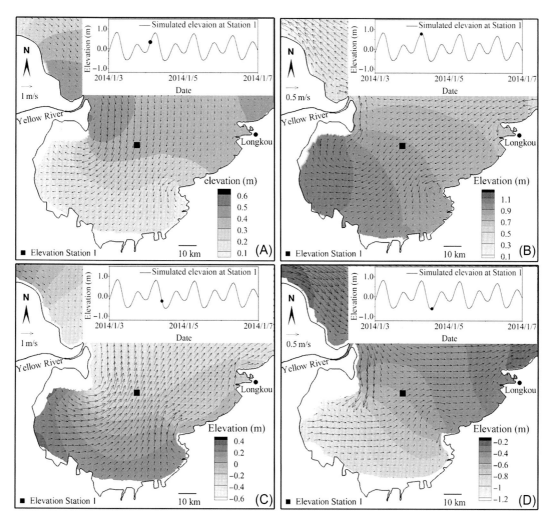

Fig. 6 Tidal surface elevations and currents in Laizhou Bay are numerically computed at four tidal phases with FVCOM: (A) rising tide, (B) high tide, (C) falling tide, and (D) low tide.

Strong flood tidal currents are off the Yellow River Delta, which is consistent with field observations. Fig. 6B shows the numerical results when the surface elevation reaches its maximum value. The computed surface elevation is shown to increase from off-shore to the nearshore region, and the maximum elevation is found in the southwest corner of the bay, attributable primarily to the local topography and the Coriolis force. In Fig. 6C, the ebb tidal current inside the bay reaches its peak value and flows northeastward, and the ebb tidal currents are shown to be stronger than the flood tidal currents. As the elevation at Station 1 reaches its minimum value, there are southward inflows off the Yellow River mouth and weak eastward currents in the northeast of the bay, as shown in Fig. 6D.

3.2 Coastal Waves

Coastal waves are main driver affecting the spatial distribution of suspended sediment and the transportation of nutrients in Laizhou Bay. In general, locally generated wind waves account for >80% of ocean waves in Laizhou Bay. As the local climate is mainly influenced by the East Asian monsoon, northerly winds prevail in winter, and relatively weaker southerly winds dominate in summer. Correspondingly, the dominant waves are from N to NNE in winter, and southerly small waves often occur in summer.

Table 1 lists mean wave height $\overline{H}$, mean wave period $\overline{T}$, and maximum wave height H_{max}, and the frequency distribution f at different wave direction, which were observed from Aril 1981 to May 1982 at a water depth of about 7.6 m at Sanshan Island Station. The dominant wave direction is NNE, consistent with the numerical results of Lv et al. (2014), and the NNE maximum wave height is 3.9 m. The measured mean wave periods are generally shorter than 5.0 s, indicating that the wind waves prevail in the bay. Short-term statistical wave heights and periods can be derived with the methods of You (2009a, 2009b).

The frequency distribution of waves in each direction is also analyzed from the same wave data collected at Sanshan Island Station and shown in Fig. 7. The waves propagating from NNE direction prevail in every season of the year. In autumn and winter, the waves from northwest, north, and northeast account for substantial portions. There are a greater number of calm days

Table 1 Wave parameters and frequency distribution observed at Sanshan Island Station

θ	N	NNE	NE	SSW	WSW	W	WNW	NW	NNW
$\overline{H}$(m)	1.2	1.4	1.0	0.5	0.7	1.0	1.2	1.3	1.0
$\overline{T}$(s)	4.7	5.0	4.1	3.5	3.7	4.3	4.6	4.7	4.7
H_{max}(m)	3.2	3.9	2.4	0.8	0.9	1.1	1.7	1.6	1.1
f(%)	5.75	11.25	2.00	0.25	0.25	0.50	2.75	6.25	4.00

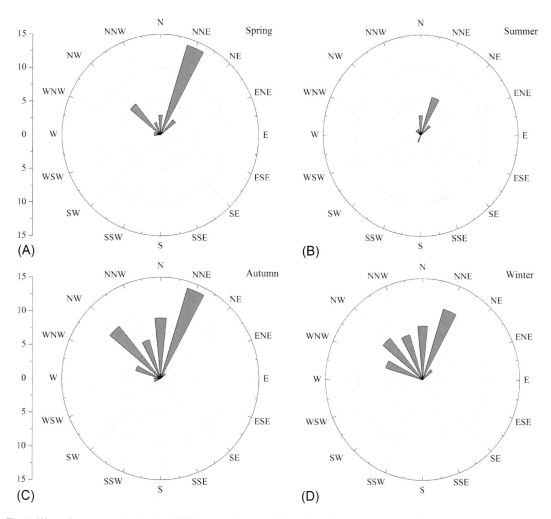

Fig. 7 Wave frequency distribution (%) observed seasonally at Sanshan Island Station. (A) Spring, (B) summer, (C) autumn, (D) winter.

in summer than in the other seasons, because relatively weaker southerly winds prevail in summer and the shorter effective wind fetch in the south of Sanshan Island.

Numerical wave models were also applied to study wave characteristics in Laizhou Bay in the last two decades. The wave model SWAN was used to numerically compute the wave characteristic in the Bohai Sea from 1985 to 2004, from which 100-year return significant wave height was extrapolated to be $H_{100} = 5.06$ m in the middle of Laizhou Bay (Wang et al., 2012). You and Nielsen (2013), You et al. (2015), and You and Yin (2016) discussed a range

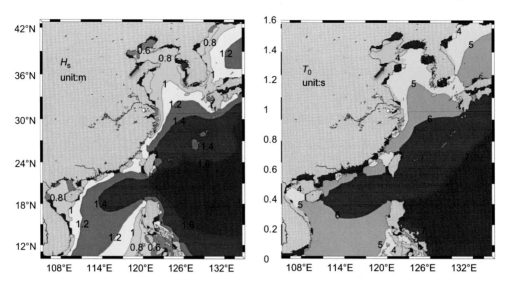

Fig. 8 Spatial variations of significant wave height and period on the coast of China averaged from 26-year time series of ERA-Interim wave reanalysis data (1990–2015), where Laizhou Bay is located at the *red arrow* point, and $H_s \approx 0.6$ m and $T_0 \approx 4$ s.

of uncertainty in estimating return wave heights from a short wave record. In analyzing 26-year time series of ERA-Interim wave reanalysis data (1990–2015), Fig. 8 depicts the spatial variations of significant wave height H_s in deep water along the coast of China (You, 2017). The wave height H_s tends to decrease spatially from the south to north on the coast, where Laizhou Bay is located at the red arrow point, and the yearly mean wave height and period are $H_s \approx 0.6$ m and $T_0 \approx 4$ s inside the bay.

4 Sediment Transport

4.1 Sediment Type

The spatial distribution of sediment types in Laizhou Bay is shown in Fig. 9, which is reanalyzed from the field data of Jiang et al. (2004) and Wang et al. (2014). The Wentworth scale is used to classify the sediment types

$$D = D_0 2^{-\phi}, \tag{1}$$

in which ϕ is the Wentworth scale and $D_0 = 1$ mm is the reference diameter to make the sediment diameter D have the same dimension as D_0. In Eq. (1), the ϕ value of 8–10 is defined for clay,

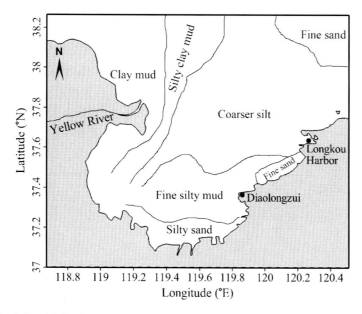

Fig. 9 Spatial distribution of sediment types in Laizhou Bay.

4–8 for silt, 3–4 for very fine sand, 2–3 for fine sand, and 1–2 for coarse sand.

In the west of Laizhou Bay and in the region close to the Yellow River mouth, sediment is clay mud, and generally finer than coarse silt and silt mud in the east of the bay. In the south of Laizhou Bay, the sediment is primarily composed of silty sand.

4.2 Sediment Settling Velocity

Cohesive sediment settling velocity w_s in Laizhou Bay is a function of suspended sediment concentration (SSC) especially in the region close to the Yellow River mouth where SSC is quite high. Before the Water Sediment Regulation Scheme (WSRS) was begun in 2002, the mean SSC at the Yellow River estuary was about 25 kg/m^3 (Wright and Nittrouer, 1995) and the maximum SSC was higher than 200 kg/m^3 (Ren and Shi, 1986). Most of the suspended fine sediments in the Yellow River were transported into the Bohai Sea in flood seasons of July to October, and about 30%–40% of sediments deposited off the Yellow River estuary.

The variation of cohesive sediment settling velocity w_s with suspended sediment concentration C in quiescent water in Fig. 10 was measured experimentally by You (2004), where $\triangle t'$ is time interval and h is water depth. The measured settling

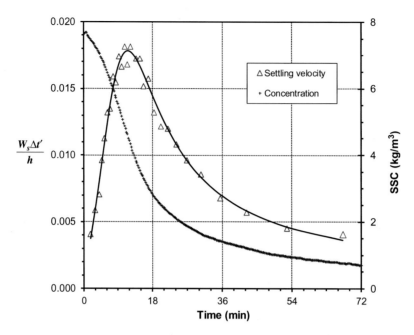

Fig. 10 Variations of cohesive sediment settling velocity w_s and suspended sediment concentration (SSC) in quiescent water with time (You, 2004).

velocity w_s was found to be independent of C in the free settling regime of $C < 0.3$ kg/m³, and then to increase nonlinearly with C in the enhanced settling regime of $0.3 \text{kg/m}^3 \leq C < 4.3 \text{kg/m}^3$, and finally to decrease sharply with C in the hindered settling regime of $C \geq 4.3$ kg/m³. The maximum settling velocity occurs at $C = 4.3$ kg/m³ and is about 9 times faster than the free-settling velocity w_0. A single empirical formula of You (2004) was proposed for the calculation of w_s at different value of C

$$w_s = w_0 \, e^{\left(0.9779C - 0.1080C^2\right)}, \tag{2}$$

in which C is measured in kg/m³, and w_0 is the sediment free settling velocity when $C = 0$.

4.3 Sediment Resuspension

Suspended sediment concentration (SSC) in Laizhou Bay is affected significantly by coastal waves and the amount of suspended fine sediments delivered by the Yellow River into the Bao-hai Sea. Fig. 11 shows the spatial variation of sea surface turbidity derived from Landsat remote images in the Yellow River Delta

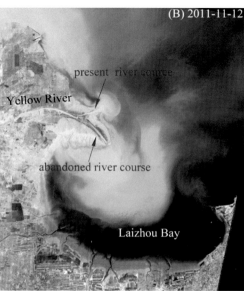

Fig. 11 Surface turbidity derived from Landsat satellite sensing images in Laizhou Bay.

region. The surface turbidity in the flood season of 1992 is shown in Fig. 11A. The sediment plume was limited to the region of the Yellow River mouth, and the SSC decreased spatially with the increasing distance from the river mouth when small wind-waves prevailed in the summer. In contrast, the spatial distribution of high SSC is presented in Fig. 11B when large wind-waves associated with strong northerly winds prevailed in winter.

Cohesive sediment resuspension depends not only on the magnitude of applied bed shear stress, but also on the bed sediment properties such as sediment rheology. For a given applied shear stress, for example, less sediment is resuspended from a fully consolidated bed than from a freshly deposited one. The erodibility of the bed of cohesive sediment is determined by a variety of physical, chemical, and biological factors (Berlamont et al., 1993). Many laboratory studies have been carried out to investigate the bed erodibility in different flumes. The main weakness of the laboratory studies is that the physical and biological properties of the bed sediment sample may not be maintained during transportation of the sediment sample from the field to laboratory. In order to overcome this weakness, different in situ benthic flumes have been developed to directly measure critical bed shear stress for sediment resuspension in the field (Amos et al., 1992; Maa et al., 1993). The in situ benthic flumes are only applicable to unidirectional flow induced by tides, not to oscillatory flow induced by waves (Fig. 11).

A comprehensive field study was undertaken by You (2005a,b) to investigate sediment resuspension dynamics in Moreton Bay, a large semienclosed bay situated in southeast Queensland in Australia. Four field deployments were carried out in the bay, two in deep water of 15–16 m (Sites 1 and 2) and the other two in shallow water of 6 m (Sites 4 and 5). The bed sediment was found to be fine sand of $d_{50} = 0.2$ mm at Site 1, and cohesive sediment at the other three sites. The field data on tidal currents, waves, suspended sediment concentrations were measured at each site with an instrumented tripod sitting on the bed. In analyzing the collected field data, the dominant driving forces for sediment resuspension in the bay were found to be location-dependent. Short wind-waves generated locally within the bay had played an important part in sediment resuspension processes in the shallow water of Site 4 and Site 5, especially during the local storm events, and the critical bed shear stress τ_{cr} for clay mud resuspension was estimated as $\tau_{cr} = 0.083 \sim 0.095$ Pa.

Sediment resuspension occurs when the applied bed shear stress τ_w is larger than the critical shear stress τ_{cr}. The bed shear stress under waves is calculated as

$$\frac{\tau_w}{\rho} = \frac{1}{2} f_w U_s^2 \qquad (3)$$

in which f_w is the wave friction factor, ρ the water density, and U_s the significant wave orbital velocity magnitude close to the seabed. There are many formulas derived for the calculation of f_w (You et al., 1991, You et al., 1992; You, 1994), but most of them are only valid for turbulent wave boundary layer flow, not proposed for transitional flow regime. The wave friction factors, which are derived from the laboratory and field data on sediment initial motion under waves, are compared with those calculated from the formulae of Jonsson (1980).

$$f_w = \exp\left[5.213\left(\frac{A}{k_s}\right)^{-0.194} - 5.977\right] \qquad (4)$$

in which $A\omega = U_s$ and ω is the angular frequency, and k_s the seabed roughness length (You, 2005a,b). The wave friction factors in Fig. 12 are estimated under the assumption that the Shields curve derived for sediment initiation in unidirectional flow is also valid under oscillatory flow induced by waves (You, 2000; You and Yin, 2006).

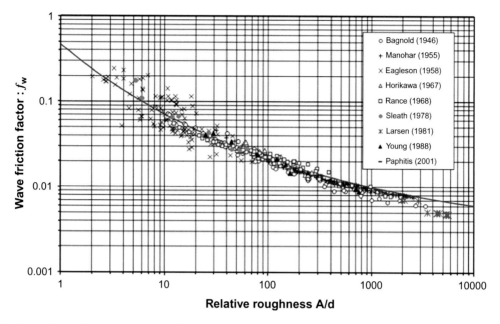

Fig. 12 Wave friction factor derived from the data on sediment initial motion under waves.

5 Human Impact

5.1 Coastal Land Reclamation

The natural coastline in Laizhou Bay has been remarkably modified primarily due to the large-scale coastal land reclamation that was carried out in recent years. These human interventions have led to pronounced alternations on local hydrodynamics and sediment transport (Pelling et al., 2013; Ding and Wei, 2017) and serious impacts on coastal ecosystems (Wang et al., 2014). Fig. 13 shows the shoreline changes in Laizhou Bay from 1997 to 2017, where large-scale land reclamation was undertaken in Longkou, Laizhou, and Weifang during the past two decades.

From 2000 to 2013, the artificial shoreline length in Laizhou Bay was found to increase by 135 km, and the reclaimed coastal land area by 372 km^2, but the natural shoreline length is about 32 km shorter. The total shoreline length of the bay is about 300 km, and the water area of the bay is about 700 m^2; this means that about 45% of the original coastline has been modified, and about 5% of the water area in Laizhou Bay has been reduced due to coastal land reclamation.

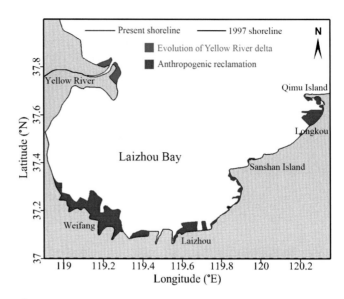

Fig. 13 Shoreline evolution and land reclamation in Laizhou Bay from 1997 to 2017.

5.2 River Reservoir Construction

Sediment transport in Laizhou Bay is significantly affected by the morphological evolution of the Yellow River Delta, and the amount of suspended sediment load delivered by the Yellow River into the bay. The coastlines of the modern Yellow River Delta in 1855 and 1934, along with the present shoreline, are shown in Fig. 14, together with historical changes of the river entrance channel. The surface area of the Yellow River Delta increased approximately 248 km^2, and the length of the coastline increased by 36 km from 1983 to 2011 (Kong et al., 2015). Since 1855, the river entrance channel has changed its course 11 times, either naturally or artificially (Zhang et al. 2016), of which four main alterations of the river entrance in the past 65 years are shown in Fig. 14. The entrance channel of the Yellow River was artificially altered to mitigate flood problems in 1964. Then, the river entrance channel was artificially shifted to the Qingshuigou River because of developing the Shengli Oilfield off the Yellow River Delta in May 1976. Finally, in August 1996, the river channel was altered to the present location to facilitate another two oilfields off the Yellow River Delta (Kong et al., 2015).

The annual river flow discharge and sediment flux in the Yellow River, which were measured from 1950 to 2015 at Lijin hydrologic station, are shown in Fig. 15 after four major water reservoirs were built in the upper steam of the Yellow River. The Lijin hydrologic station is located 100 km upstream of the Yellow River mouth. The

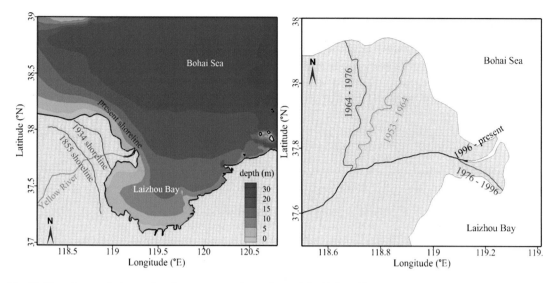

Fig. 14 Shoreline evolution and major entrance alternations of the Yellow River.

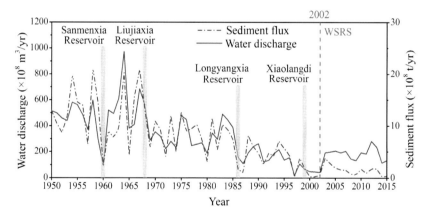

Fig. 15 Annual water discharge and sediment flux measured from 1950 to 2015 at Lijin hydrologic Station 100 km upstream of the Yellow River's mouth.

Sanmenxia, Liujiaxia, and Longyangxia reservoirs were constructed in 1960, 1968, and 1986 respectively. Consequently, the water discharge became about 4 times smaller from $588 \times 10^8 \, \text{m}^3/\text{yr}$ to $157 \times 10^8 \, \text{m}^3/\text{yr}$ and the sediment load was about 7 times smaller from $13.2 \times 10^8 \, \text{t/yr}$ to $1.69 \times 10^8 \, \text{t/yr}$. After the Xiaolangdi Reservoir was completed in 1999, the water discharge further reduced to $66 \times 10^8 \, \text{m}^3/\text{yr}$, which only accounts for 12% of the river flow discharge in 1950, and the sediment load reduced to $1.87 \times 10^8 \, \text{t/yr}$, about 14% of the sediment load in 1950.

References

Amos, C.L., Grant, J., Dborn, G.R., Black, K., 1992. Sea carousel—a benthic, annular flume. Estuar. Coast. Shelf Sci. 34, 557–577.

Berlamont, J., Ockenden, M., Toorman, E., Winterwerp, J., 1993. The characterisation of cohesive sediment properties. Coast. Eng. 21, 105–128.

Chen, C., Huang, H., Beardsley, R.C., Liu, H., Xu, Q., Cowles, G., 2007. A finite volume numerical approach for coastal ocean studies: comparison with finite difference models. J. Geophy. Res. Oceans (C3), 112. https://doi.org/10.1029/2006JC003485.

Chen, C., Huang, H., Beardsley, R.C., Xu, Q., Limeburner, R., Cowles, G.W., Lin, H., 2011. Tidal dynamics in the Gulf of Maine and New England shelf: an application of FVCOM. J. Geophys. Res. Oceans (C12), 116. https://doi.org/10.1029/2011JC007054.

Chen, C., Liu, H., Beardsley, R.C., 2003. An unstructured grid, finite-volume, three-dimensional, primitive equations ocean model: application to coastal ocean and estuaries. J. Atmos. Ocean. Technol. 20 (1), 159–186.

Ding, Y., Wei, H., 2017. Modeling the impact of land reclamation on storm surges in Bohai Sea, China. Nat. Hazards 85 (1), 559–573.

Egbert, G.D., Erofeeva, S.Y., 2002. Efficient inverse modeling of barotropic ocean tides. J. Atmos. Ocean. Technol. 19 (2), 183–204.

Galperin, B., Kantha, L.H., Hassid, S., Rosati, A., 1988. A quasi-equilibrium turbulent energy model for geophysical flows. J. Atmos. Sci. 45 (1), 55–62.

Jiang, W., Pohlmann, T., Sun, J., Starke, A., 2004. SPM transport in the Bohai Sea: field experiments and numerical modelling. J. Mar. Syst. 44 (3), 175–188.

Jonsson, I.G., 1980. A new approach to oscillatory rough turbulent boundary layers. Ocean Eng. 7, 109–152.

Kong, D., Miao, C., Borthwick, A.G., Duan, Q., Liu, H., Sun, Q., Ye, A., Di, Z., Gong, W., 2015. Evolution of the Yellow River Delta and its relationship with runoff and sediment load from 1983 to 2011. J. Hydrol. 520, 157–167.

Lv, X., Yuan, D., Ma, X., Tao, J., 2014. Wave characteristics analysis in Bohai Sea based on ECMWF wind field. Ocean Eng. 91, 159–171.

Maa, J.P.-Y., Wright, L.D., Lee, C.-H., Shannon, T.W., 1993. VIMS Sea Carousel: a field instrument for studying sediment transport. Mar. Geol. 115, 271–287.

Mellor, G.L., Yamada, T., 1982. Development of a turbulence closure model for geophysical fluid problems. Rev. Geophys. 20 (4), 851–875.

Pelling, H.E., Uehara, K., Green, J.A.M., 2013. The impact of rapid coastline changes and sea level rise on the tides in the Bohai Sea, China. J. Geophy. Res. Oceans 118 (7), 3462–3472.

Ren, M.E., Shi, Y.L., 1986. Sediment discharge of the Yellow River (China) and its effect on the sedimentation of the Bohai and the Yellow Sea. Cont. Shelf Res. 6 (6), 785–810.

Wang, W., Liu, H., Li, Y., Su, J., 2014. Development and management of land reclamation in China. Ocean Coast. Manage. 102, 415–425.

Wang, X.H., Wang, H., 2010. Tidal straining effect on the suspended sediment transport in the Huanghe (Yellow River) estuary, China. Ocean Dyn. 60 (5), 1273–1283.

Wang, Z.F., Wu, K.J., Zhou, L.M., Wu, L.Y., 2012. Wave characteristics and extreme parameters in the Bohai Sea. China Ocean Eng. 26 (2), 341–350.

Wright, L.D., Nittrouer, C.A., 1995. Dispersal of river sediments in coastal seas: six contrasting cases. Estuaries 18 (3), 494–508.

You, Z.J., Yin, B.S., 2016. In: Standardized procedure for estimation of extreme ocean waves. 26th International Ocean and Polar Engineering Conference, Rhodes, Greece, June 26–July.

You, Z.J., 2009a. A close approximation of wave dispersion relation for direct calculation of wavelength in any coastal water depth. Appl. Ocean Res. 30, 133–139.

You, Z.J., 2009b. Statistical distribution of nearbed wave orbital velocity in intermediate coastal water depth. Coast. Eng. 56, 844–852.

You, Z.J., 2017. Assessment of coastal inundation and erosion hazards along the coast of China. International Ocean and Polar Engineering Conference, San Francisco, California, 25–30 June.

You, Z.J., Yin, B.S., Ji, Z.Z., Hu, C., 2015. Minimization of the uncertainty in estimation of extreme waves. J. Coast. Res. 75, 1277–1281.

You, Z.J., Nielsen, P., 2013. Extreme coastal waves, ocean surges and wave runup. In: Finkl, C.W. (Ed.), Coastal Hazard Book. In: Coastal Research Library, vol. 6. Springer, The Netherlands. https://doi.org/10.1007/978-94-007-5234-4-22 Chapter 22).

You, Z.J., 2000. A simple model of sediment initiation under waves. Coast. Eng. 41, 399–412.

You, Z.J., Nielsen, P., Wilkinson, D.L., 1991. Velocity distributions of waves and currents in combined flows. Coast. Eng. 15, 525–543.

You, Z.J., 2005a. A field study of fine sediment resuspension dynamics in a large semi-enclosed bay. Ocean Eng. 32, 1982–1993.

You, Z.J., Yin, B.S., 2006. A unified criterion for initiation of sediment motion and inception of sheet flow under waves. Sedimentology 53, 1181–1190.

You, Z.J., 2004. The effect of suspended sediment concentration on the settling velocity of cohesive sediment in quiescent water. Ocean Eng. 31, 1955–1965.

You, Z.J., 2005b. Estimation of bed roughness from mean velocities measured at two levels near the seabed. Cont. Shelf Res. 25, 1043–1051.

You, Z.J., Nielsen, P., Wilkinson, D.L., 1992. Velocity distribution in turbulent oscillatory boundary layer. Coast. Eng. 18, 21–38.

You, Z.J., 1994. A simple model for current velocity profiles in combined wave-current flows. Coast. Eng. 23, 289–304.

Yu, Y., Wang, H., Shi, X., Ran, X., Cui, T., Qiao, S., Liu, Y., 2013. New discharge regime of the Huanghe (Yellow River): causes and implications. Cont. Shelf Res. 69, 62–72.

Zhang, H., Chen, X., Luo, Y., 2016. An overview of ecohydrology of the Yellow River delta wetland. Ecohydrol. Hydrobiol. 16 (1), 39–44.

Further Reading

Bagnold, R.A., 1946. Motion of waves in shallow water: interaction between waves and sand bottom. Proc. Roy. Soc. Lond. 187, 1–15.

Dingler, J.R., Inman, D.L., 1975. Wave formed ripples in seashore sands. Proceedings of 15th Coastal Engineering Conference ASCE, 2109–2126.

Eagleson, P.S., Dean, R.G., Peralta, L.A., 1958. The Mechanics of the Motion of Discrete Spherical Bottom Sediment Particles due to Shoaling Waves. Technical Memo, No. 104,U.S. Army Corps of Engineers, Beach Erosion Board.

Larsen, L.H., Sternberg, R.W., Shi, N.C., Marsden, M.A.H., Thomas, L., 1981. Field investigations of the threshold of grain motion by ocean waves and currents. Mar. Geol. 42, 105–132.

Lavelle, J.W., Mofjeld, H.O., Baker, E.T., 1984. An in situ erosion rate for a fine-grained marine sediment. J. Geophys. Res. 89, 6543–6552.

Madsen, O.S., Grant, W.D., 1975. The threshold of sediment movement under oscillatory waves. A discussion. J Sediment. Petrol. 45, 360–361.

Manohar, M., 1955. Mechanics of Bottom Sediment Movement Due to Wave Action. Technical Memo. No. 75, U.S Army Corps of Engineers, Beach Erosion Board.

Marine Geology Laboratory of Institute of Oceanology Chinese Academy of Sciences, 1985. The Bohai Sea Geology. Science Press, Beijing, pp. 1–232 (in Chinese).

Mehta, A.J., 1988. Laboratory studies of cohesive sediment deposition and erosion. In: Physical Processes in Estuaries. Springer Verlag, Berlin, pp. 427–445.

Paphitis, D., Velegrakis, A.F., Collins, M.B., Muirhead, A., 2001. Laboratory investigations into the threshold of movement of natural sand-sized sediments under unidirectional, oscillatory and combined flows. Sedimentology 48, 645–659.

Rance, P.J., Warren, N.F., 1968. The threshold of movement of coarse material in oscillatory flow. Proceedings of 11th Conference on Coastal Engineering, pp. 487–491.

Roman, M.R., Tenore, K.R., 1978. Tidal resuspension in Buzzards Bay, Massachusetts I. Seasonal changes in the resuspension of organic carbon and chlorphyll a. Estuar. Coast. Shelf Sci. 6, 37–46.

Sleath, J.F.A., 1978. Measurements of bed load in oscillatory flow. J. Waterway Port Coast. Ocean Eng. ASCE 104, 291–307.

You, Z.J., Yin, B.S., 2007. Direct measurement of bed shear stress under waves. J. Coast. Res. 50, 1132–1136.

Young, J.S.L., Sleath, J.F.A., 1988. In: Initial motion in combined wave and current flows. Proceedings of 21st Conference on Coastal Engineering, Malaga, ASCE, New York, pp. 1140–1151.

8

REMARKS

Xiao Hua Wang, Isabel Jalón-Rojas

The Sino-Australian Research Centre for Coastal Management, The University of New South Wales, Canberra, ACT, Australia

In this book, we aim to examine and document the physical and ecological impact on estuarine and coastal environment due to intensive human activities and their effects on the coastal ecosystems in China. A good understanding of the current state of these marine environment and lessons learned from these human influences would be extremely valuable to restore and protect these habitats and ecosystems from further environmental degradation and catastrophe. A total of six case studies by the authors collectively tell a story about how these coastal environments respond dynamically to severe human-induced perturbations, and where things have gone right or wrong from environmental and resource-management points of view. Admittedly, many of the impacts of these human interventions on the estuarine environment and their processes are still not well understood and are subject to ongoing studies. Based on what we have learned, suggestions and recommendations on how we document and address these issues are offered, at least in some of these study sites, to remediate the problems and prevent future failure in other world estuaries.

In Chapter 2, Lulu Qiao and colleagues describe the environmental evolution of Jiaozhou Bay due to heavy human intervention through field observations, satellite observations, and numerical model simulation.

The replacement of natural tidal flats by salt ponds, aquaculture area, and reclamation for ports decreased the sea area and increased the artificial coastal line over the last seven decades. As a consequence, the maximum velocity speed was reduced, the anticyclonic eddy at the bay mouth decreased dramatically, the M_2 tidal-energy flux across the entrance was attenuated, and the tidal prism was reduced. This implied the decrease of

Sediment Dynamics of Chinese Muddy Coasts and Estuaries. https://doi.org/10.1016/B978-0-12-811977-8.00008-X

the bay erosion, and therefore, of suspended sediment concentrations, and the decline of the water exchange capability, and therefore, the accumulation of nutrients and pollutants. The water exchange rate was further decreased due to a sea-cross bridge construction in 2011, which also resulted in coarse sediments in the north side of the bridge. To all this must be added the dramatic increase of inorganic dissolved nitrogen and of active phosphate because of the population growth and the economic development around the bay. Consequently, water quality has been dramatically degraded, and environmental issues such as eutrophication were aggravated, triggering algal blooms such as green tides.

The biological system of Jiaozhou Bay also varied in relation to these pressures. The production function, regulatory function, and biological diversity of the bay have decreased during the last three decades. While phytoplankton and fish biodiversity significantly decreased, the zooplankton biodiversity increased. Management recommendations and politics such as real-time and long-term monitoring and Integrated Coastal Zone Management (ICZM) are finally discussed.

In Chapter 3, Jiabi Du and collaborators summarize the physical, ecological, and anthropogenic processes in the muddy coast of Jiangsu. The 90% of the 954 km Jiangsu coast are covered by intertidal mudflats which are developed and controlled by the following factors: (1) a net landward sediment transport induced by the convergence of two tidal waves, a strong flood-dominant tidal asymmetry, and a low energy environment; (2) large sediment sources from the ancient Yellow River delta, the Yangtze River, and the offshore sand ridge; (3) the trapping of sediments induced by salt marshes. In particular, the introduction of *Spartina anterniflora* increased the accumulation of fine sediment and the generation of new land, although it caused ecological problems such as the loss of biodiversity, the alteration of ecosystems processes, and the degradation of native habitats.

Human intervention has affected the evolution of mudflat in the Jiangsu Coast. Reclamation activities have been ongoing from 2005, decreasing the accumulation rate of sediment, and therefore, the mudflat width. The seasonal installation of aquaculture infrastructures weakened hydrodynamics increasing bottom sediment and the frequency of accretion events. Together with the accelerated sea-level rise, coastal protection engineering infrastructures and damming in local rivers are decreasing the supply of sediment, limiting the seaward progress of mudflat.

In Chapter 4, Jianrong Zhu and colleagues describe and analyze the hydrodynamics, saltwater intrusion, and hypoxia in the Changjiang Estuary by using ECOM-si model. River discharge, tide, wind, mixing, topography, and continental shelf current out-side the river mouth control the hydrodynamic processes, which are strongly variable over time and space. The net landward flow in the North Branch during low river flow generated saltwater intrusion. The increase of water level in upper reaches during spring tides leads to a large saline water spilling over the shoal into the South Branch. Salinity intrusion is enhanced by strong north-erly winds and sea level rise, and impacted by human interven-tions such as the deep navigation channel project and the Three Gorges Reservoir. The deep navigation channel project decreased saltwater intrusion at the North Channel and the lower North Passage, while increasing it at the upper North Passage. The Three Gorges Reservoir supplies freshwater input during dry season, suppressing salt intrusion.

Hypoxia can occur at the Changjiang mouth due to three major causes. First of these is the oxygen depletion due to decomposi-tion of organic matter from algae. The fast development of agriculture and industry in the Changjiang Basin has increased the terrestrial inputs of nutrients and pollutants in the estuary over time. This increase has enhanced eutrophication and led to frequent algal bloom events. Second is the obstruction of vertical oxygen exchange by the pycnocline, which is intensified during periods of high river discharge. Third is the low current speed at the bottom, which can dramatically decrease dissolved oxygen and favors the maintenance of hypoxia.

In Chapter 5, Li Li and collaborators analyze the impact of large-scale land reclamation on the physical environment of Hangzhou Bay over the last six decades. Coastline changes are quantified from Landsat images, showing a decrease of the estuary width up to 5 km between 1962 and 2015.

The consequences of these morphological changes on coastal hydrodynamics are analyzed using numerical modeling. The decrease of the estuary width increased the tidal choking effect, but even more the shoaling effect, amplifying tidal range at the upper reaches. The tidal prism was reduced at middle reaches, and therefore, tidal currents and tidal directions were modified at the upper estuary from almost west-east direction to the northwest-southeast direction. The seaward residual currents increased from 1974 to 2016 at the upper estuary but decreased at lower reaches.

Geometrical changes also narrowed the main tidal channel and decreased its radius of curvature. Hence, the centrifugal force increased and the tidal channel migrated northward. This may enhance the deposit of sediments inside the bay, impacting nav-igation channels and coastal stability. A further narrowing of the sections of Hangzhou Bay might increase tidal choking, damping its popular tidal bore, thus, negatively affecting tidal bore tourism.

In Chapter 6, Keliang Chen and collaborators describe the climatology, hydrology, geology, and biodiversity of Xiamen Bay and evaluate the impacts of nutrient pollution, reclamation, and dredging. The nutrient input from the Jiulong River has recently increased, leading to eutrophication and frequent red tides that have changed the structure of the plankton community. From 1955, Xiamen Bay has been subject to several engineering works such as dikes and dams in order to increase land resources and port infrastructures. As a consequence, the tidal prism decreased, and so have the current velocity and the sediment-carrying capability. These changes implied siltation problems, water quality degradation, the modification of habitats, and the losses of biodiversity and ecosystem services. In recent years, extensive dredging has been implemented to support the coastal reclamation project. Tides drive the dredging-produced sediment to protection zones of *Branchiostoma* special and dolphin natural reserve, damaging these ecosystems.

Finally, authors provide recommendations to recover the good ecological status of the bay, including the development of laws and regulations, scientific planning, strategic environmental assessment, and the improvement of the environmental capacity of the bay by mitigating human interventions.

In Chapter 7, Zai-Jin You and Chen Zhao review the physical oceanography of Laizhou Bay, including the meteorological climate, tides, waves, and sediment transport, and evaluate the impact of land reclamation and river reservoir constructions. Laizhou Bay is subject to the East Asian Monsoon and is affected by cold-air outbreaks and tropical cyclones, which lead to major coastal hazards and important economic losses. Suspended sediment concentrations mainly depend on locally generated wind waves and the sediment input of the Yellow River.

Massive reclamation has increased coastal area by $372\,km^2$ from 2002 and 2013. This implies that 45% of the original coastline has been modified and about 5% of the water area has been decreased. The authors claim that these human interventions have led to pronounced alterations on local hydrodynamics and sediment transport, as well as serious impact on coastal ecosystems.

The discharge and sediment inputs of the Yellow River dramatically decrease after the construction of four major water reservoirs in 1960, 1968, 1986, and 1999. The impact of these human interventions on the estuarine environment is still not well understood and is subject to further studies.

INDEX

Note: Page numbers followed by *f* indicate figures, and *t* indicate tables.

A

Anthropogenic pressure
    indicators (APIs), 15
Aquaculture, 39–40

B

Bohai Sea, Yellow Sea, and East
    China Sea (BYECS), 40–41

C

Changjiang Estuary
    coastal hypoxia, 67–71, 69–70*f*,
        145
    river discharge, 55, 55*t*
    saltwater intrusion
        deep navigation channel
            project, 144–145
        sea level rise, 56*f*, 64–65
        waterway project impact,
            65–66
        wind, 64
    TGR project, 66, 67*f*
    water movement
        residual current, 60–62,
            63*f*
        runoff, 55–56
        tide, 56–57, 58–59*f*
        wind, 57–60, 60–62*f*
Changjiang River, 83
China Mangrove Conservation
    Network (CMCN), 117
China's Blue Economic Strategy,
    18
China's coastal ecosystems
    biophysiochemical processes,
        2, 3*f*
    global climate change, 1
    human pressures, 1
    study sites, location map
        of, 3*f*

Chinese National Seawater
    Quality Standard
    (CNSQS), 9–11
Coastal carrying indicators
    (CCIs), 15
Coastal land reclamation
    HZB, 78–80
    Jiangsu mudflat, 37–39
    JZB, 12–15
    Laizhou Bay, 137, 138*f*
    Xiamen Bay, 104–105

D

Dafeng Port, 41
Dagang tide-gauge station, 6–8
Dagu River, 8
Damming, 40–41
Datong Hydrographic Station,
    55–56

E

Ecological disasters
    JZB, 16–17
    Xiamen Bay, 16–17
Ecological footprint (EF)
    method, 107
ECOM-si model, 53, 54*f*, 144–145
Environmental Fluid Dynamics
    Code (EFDC), 108–109
Eutrophication index (EI),
    16–17

F

Finite volume coastal ocean
    model (FVCOM), 13,
    82–83, 128

G

Grain-size trend analysis (GSTA)
    method, 112

H

Haixi-Qianwan Bay, 9
Halophytic vegetation, 34
Hangzhou Bay (HZB)
    coastal engineering
        stability, 90
        tidal choking effect, 90
    coastline changes, 78–80, 80*f*,
        145–146
    field data analysis, 82
    flood currents, 78
    location map, 78, 79*f*
    physical environment, 82–89
    remote sensing data analysis,
        80, 81*t*
    tides changes, numerical
        study of
        model setup and validation,
            82–83, 84*f*
        residual currents, 88–89, 89*f*
        tidal channel changes,
            86–88, 87–88*f*
        tidal range, 84–85, 85*f*
        vertically averaged tidal
            current ellipses, 85–86, 86*f*
    urbanization, 78–81, 91
Huangdao Port, 11
Human interventions
    Jiangsu mudflat
        aquaculture, 39–40, 144
        damming, 40–41
        reclamation, 37–39, 144
        shoreline protection, 40, 144
    JZB
        biological system, 15–16
        coastline change, 12–13
        ecological disasters, 16–17
        hydrodynamics, 13
        sediment transport, 13–14
        water quality, 14–15

Human interventions
(Continued)
Laizhou Bay
coastal land reclamation,
137, 138*f*
river reservoir construction,
138–139
Xiamen Bay
comprehensive law
enforcement system, 117
environmental impacts,
105–108
reclamation, 104–105
scientific planning, 117
Hydrodynamics
Jiangsu mudflats, 28–30
JZB, 6–8

I
Intensive dredging, 41
Intertidal mudflat, 26, 34

J
Jiangsu mudflats
biogeochemical processes,
43–44
controlling factors, 144
ecological control, 34–37
hydrodynamics, 28–30
location map, 26, 27*f*
numerical modeling, 42–43
saltmarsh distribution, 34–36,
35*f*
sediment source, 30–32
sediment transport, 32–33
storm's impact, 43
Jiaozhou Bay (JZB)
biological system, 11–12, 144
coastline and sea area change,
6–8, 7*f*, 12–13
Hongdao Channel, 18
hydrodynamics, 6–8
land-based wastewater
heavy metal contamination,
10–11, 10*f*
inorganic dissolved nitrogen
(DIN), 9–10, 10*f*, 15
persistent organic
pollutants (POPs), 11

petroleum, 11
Qingdao Municipal
Government, 18–19
rivers, 8
seabed morphology, 6
sea-cross bridge, 17–18,
143–144
sediment transport, 8–9
surface area, 6
water quality, 9–11
waves, 8
Jiaozhou Bay Coastal Wetland
Marine Special Reserve,
18–19
Jiulong River, 97–98, 103–104
JZB. *See* Jiaozhou Bay (JZB)

L
Laizhou Bay
astronomic tides, 127–130
coastal dynamics, 127–132
coastal waves, 130–132
cohesive sediment setting
velocity *vs.* suspended
sediment concentration,
133–134, 134*f*
cold-air outbreaks, 126–127,
127*f*
ERA-Interim wave reanalysis
data, 131–132, 132*f*
location map, 123, 124*f*
meteorological climate,
125–127
overall rainfall, 125, 125*f*
physical oceanography, 146
sediment resuspension,
134–136
sediment types, spatial
distribution of, 132–133,
133*f*
tidal currents and elevations,
computational domain of,
128–130, 128–129*f*
tropical cyclones, 126–127
wind directions, frequency
distribution of, 126,
126*f*
Lijin hydrologic station, 138–139,
139*f*

M
MIKE21 hydrodynamic model,
107–108

N
NaoTide dataset, 54

P
Porphyra cultivation, 39–40
Princeton Ocean Model (POM),
53, 107–108

Q
Qiantang River, 78, 82–83
Qingdao Port, 6–8, 11

S
Saltwater-spilling-over (SSO),
62–64
Sanshan Island Station, 130–131,
130*t*, 131*f*
Sediment transport
Jiangsu mudflats, 32–33
JZB, 8–9
Laizhou Bay, 132–136
Shoreline protection
engineering, 40
Spartina anglica, 34–36, 35*f*
Spartina anterniflora, 34–37,
144
State Environmental Protection
Administration of China
(CSEPA), 11
State Oceanic Administration
of China (SOA),
105–106
Suspended sediment
concentration (SSC)
Jiangsu mudflats, 30
JZB, 8, 9*f*
Laizhou Bay, 133–134,
134*f*
Xiamen Bay
dredging, 114–116,
115–116*f*
temporal and spatial
variation, 112–114,
113–114*f*
SWAN wave model, 55*t*, 60

T

Third Institute of Oceanography
(TIO), 105–106
Tongan Bay, 106–108
Total Allowable Area for Coastal
Reclamation (TAACR),
107–108

U

United States Environmental
Protection Agency
(US EPA), 11
University of New South Wales
Sediment model
(UNSW-Sed), 13

W

Water Sediment Regulation
Scheme (WSRS), 133

Wentworth scale, 132–133
Wild Animal Conservation Law
of the People's Republic of
China, 100–101

X

Xiamen Bay
biodiversity, 100–101
cleaner production
technology, 119
coastline changes, 104,
105*f*
ecology diversity, 102
environmental capacity,
118
geographical barriers, 100
location map, 96*f*
numerical model, 109–112
reclamation, 146

sediment concentration,
109–116
species, 101
statistical wind data, 99–100
tides, 98
topography, 98–99
Xiamen government,
117–120
Xiaolangdi Reservoir, 138–139
Xiuzhen River, 26

Y

Yanghe River, 6
Yangtze River, 26, 31, 40–41,
51–52
Yellow River, 31–32, 138–139
Yellow River Delta, 31–32,
138
Yellow Sea, 28–30, 32, 39